Progressivism:
Our Road to Serfdom

Arise America: Rebuild Your God-Given
Capitalist Foundations

Zester and Marilyn J. Hatfield

Order this book online at www.trafford.com
or email orders@trafford.com

Most Trafford titles are also available at major online book retailers.

Printed in Victoria, BC, Canada.

ISBN: 978-1-4269-3386-8 (sc)
ISBN: 978-1-4269-3387-5 (dj)
ISBN: 978-1-4269-3388-2 (e-b)

Library of Congress Control Number: 2010907264

*Our mission is to efficiently provide the world's finest, most comprehensive book publishing
service, enabling every author to experience success. To find out how to publish your book, your
way, and have it available worldwide, visit us online at www.trafford.com*

Trafford rev. 05/21/2010

Trafford
PUBLISHING® www.trafford.com

North America & international
toll-free: 1 888 232 4444 (USA & Canada)
phone: 250 383 6864 ♦ fax: 812 355 4082

Dedicated to America's

Entrepreneurs, Business Owners and
Their Employees

Books by Zester and Marilyn Hatfield

Job Security In a High-tech World

Knights In Shining Armor

Daddy's Little Girl and Mommy's Little Boy

Progressivism: Our Road To Serfdom

Introduction

Americans today are accused of having raped the world because we consume so much of the world's goods and services. Our self confidence is perhaps at an all-time low, and the American dream is said to be dead.

Why has this happened to us, and what vision of hope do we have for the future? Should we be ashamed of our past accomplishments and hang our heads in despair? Do we still have a place in contemporary world leadership for producing the world's goods and providing a comfortable standard for our own people? Will the present organized labor struggle for job security and Full Employment provide the answer of hope?

Progressive Socialists tell wage earners that they deserve this and more. If so where is the evidence that the Progressive Socialists understand how wealth is created and how to equitably distribute it? These and many other questions of vital importance are examined in this book. We examine these questions in the light of the historic realities of who we are, where we have come from and where we are going as it relates to wages, job security, Full Employment, personal wealth accumulation and our personal freedoms.

Americans are witnessing two world-changing phenomena that appear on the surface to be unrelated but which are, in fact, intimately related. On the one hand we see the crumbling economies and social unrest of several third world countries, many of which are waging war or preparing for war. We, on the other hand, are personally on the edge of the greatest era of technological explosion that the world has ever seen. Fifth generation computers with artificial intelligence are already working in many small, medium, and large production centers with major military and commercial applications.

This is not just a peripheral technological advancement that simply reshuffles the job market and, for the most part, upsets only organized labor leaders and a few Progressive Socialists. This is the dawning of the era of "smart computers," working millions of times faster with trillions of times more capacity than those of yesterday, even to the point of "artificial thinking" and decision making without external help or, for that matter, the need for any further human input. It is the thinking and independent action capability of this new generation of computers that gives rise to the terms "smart computers," "smart bombs" and "artificial intelligence." As an example, there are now large breweries that have whole divisions that are completely automated and run by as few as three people per shift.

There is no evidence that the introduction of these smart computers and their practical applications will have any respect of persons or fields of employment. The massive job displacement, social unrest, and upheaval that characterized the Industrial Revolution of the 1800s will seem like child's play compared to the challenges that await the unsuspecting American wage earners of today. White collar, middle management, and blue collar workers alike are in danger of being swept away by this new wave of automated mentality. Many feel insecurity and personal threat due to the social unrest and wars in so many sectors of our world—but this pales in comparison to the social and economic upheavals rapidly enveloping the American worker in massive layoffs even as we write this book.

According to the "American Dream" of financial security and personal fulfillment, the "haves" are those who can save enough for a down payment on a house, a car, a good education for children, furniture, good clothes, vacations, and have a little money stashed away for a rainy day. *Good paying jobs with steady wages,* liberal medical benefits, and retirement plans are supposed to provide the financial security to make it all happen.

But today millions of American workers are losing their jobs and their benefits due to technological displacement and inadequate distribution methods. For them the American Dream is dead, and tomorrow it will be dead for millions more. In the face of such national and global instability and insecurity, it becomes apparent that millions who have depended upon the concept of job security with regular paychecks to qualify them as haves are really only the precariously employed have-nots. So what really makes a have and a have not? Without understanding who we are in the process of creating wealth we will never have a true American Dream!

In order to answer these questions and to objectively view the world scene today and discover our most positive options, it is necessary to review how we came to accept our present ideas on the creation of wealth and its distribution. Surely we could not have come this far in our understanding of many disciplines of science and technology only to fail in our attempts to understand what we should do with the fruit of our success! It's sad to say, but this is exactly what we have done!

We are a country and a people brought together and molded by a strong belief in the presence of God in our lives and in our destiny. We do not believe that we were ordained of God to be a country of have-nots, nor could we say that we believe that God has ordained that other countries should be have-nots.

Despite our convictions, great imbalances are taking place. Technological advances have improved beyond the ability of our present day methods of distribution to cope with their capacity to create new wealth. In the meantime, Obama and his Progressive Socialists are using this crisis to consolidate their hold on power and to extend it to ever new levels of total control.

Progressivism: Our Road To Serfdom reveals the reasons and the history behind our current failing economy, with step by step understanding on how to rebuild our God-given Capitalist foundations with greater success. It also takes a candid look at how new wealth is created by our current and advancing technologies. As Americans, we must take part in the distribution of this new wealth, equitably and fairly, without a Progressive Socialist government takeover. We do not need (nor can we survive) Obama and Progressive Socialists telling us how much wealth we can create and how much we can have. The disease that has been growing in the bowels of our republic for over one hundred years is Progressive Socialism.

Progressivism: Our Road To Serfdom reveals the secrets and the inner workings of the Progressive Socialism, the political and ideological barrier that stands between all of us who love America and our republic's limited form of government.

Chapters One thru Five examine many of the historical beginnings of our present views and practices on creating and distributing wealth. The archaic and incorrect assumption that labor is the only source of wealth for workers and citizens, and that it should primarily be distributed through a system of wages, is challenged. In its place is presented a concept of three basic elements as the source of all wealth—labor, creative initiative, and spiritual inspiration. These elements and their effects on our technological discoveries, national and international competition, profits and losses, the workforce and its environment are presented and dealt with as a Capital Compound Theory of Value. Our Personal Capital Compound encompasses our physical labor, creative initiative, and spiritual inspiration realities, as opposed to Karl Marx's Labor Theory of Value, which only takes into account our physical factors.

Our own Progressive Socialists are the direct political and ideological descendants of Karl Marx and these Communist theories. This section is very important as a foundation for understanding the rest of the book.

Thus, the reader is given a perspective that sweeps from the corporate boardroom to the common laborer, from the centers of power to the farthest outpost of untamed wilderness, with a sense of the ageless struggle of all human societies to create wealth and distribute it. Through this study comes hope and confidence that we do have the ability to structure our lives, to maximize our standards of living, and to share the resulting wealth without displacing or enslaving other people for our advantage.

The American Dream of financial security and personal fulfillment is much more than just a dream, for both Americans and for all free people the world over. Especially for Americans as we are protected by our republic form of limited government, authorized and protected by our constitution. By taking a firm grasp on our total wealth creating element—labor, creative initiative, and spiritual inspiration—as parts of a larger whole, we will re-examine our technological breakthroughs of the last two centuries and see how these elements dictate both the opportunities and define the method by which Americans will successfully exercise their inalienable rights in their pursuit of happiness!

In our postmodern world high technology has now tipped the wealth generating scale. Inevitably, hourly wage employment is on the losing end. Technology is so powerful that every American must become a partner with it in order to accumulate personal equity and personal wealth that replaces the hourly wage! This is a strange thought to most Americans of today, but it was the common sense understanding for the first two hundred years of the colonists. The colonists did not suffer the loss of old family ties and the dangers of crossing an unforgiving ocean in flimsy sailboats looking for a wage. If they had believed that a wage was the road to personal freedom and independence, they would have stayed in the Old World. The colonist sacrificed everything to come to America in search of their opportunity to *own* the source of their wealth. Our colonist forefathers have left us a great legacy in faith, work ethic, principled living, examples of financial independence, and finally the most powerful formula for self-governance the world has ever seen.

A bystander approached Benjamin Franklin outside of Independence Hall as he left the final meeting of the Continental Congress. The bystander asked Franklin: "Sir what kind of government do we have?"

Franklin answered: "My dear sir, you have a republic if you can keep it."

Arise America, rebuild your God-given Capitalist foundations. With God's grace and providence you have nothing to fear!

Table of Contents

Chapter One

Creating New Wealth

The central core of life and liberty is greatly impacted by financial wealth: how it is perceived, understood, created, and distributed among the citizens of the world. There is great illiteracy of such issues in the world today. This first section is the first step in healing that illiteracy.

THE DREAM

Wealth: gold, silver, jewels, land, oil wells, factories, large corporations, ships, planes, a king's ransom, and a pirate's chest—words and fables that make our minds reel with the feel of power and independence. Where does it come from? How can we all get our share? These are questions that have been asked in every land by every race for thousands of years.

Everyone wants a piece of the action, but few truly understand how to do this. Even the ones who have demonstrated their ability to amass fortunes have not successfully duplicated themselves in significant numbers. If they could have they would have, but even they find themselves amazed at their good fortune to have done so well in a world that is not known for revealing its secrets of wealth and its equitable distribution.

1

To understand this dilemma, and to deliver ourselves from the jaws of socialism, we must first identify several key elements of the whole. We must take a little time and make an effort to understand the details of economic issues that we have often ignored as boring and uninteresting. It's funny how economics gets very exciting when it is represented in the discovery of new wealth that one can call one's own. It's exciting when it is our money and our hope of fulfilling our dreams that we see laying there in our hands.

We must also learn the identity of our enemies and those who work diligently to gain control over our lives and everything that we hold dear. We can no longer afford the luxury of sitting on the sidelines and letting everyone who wants a piece of us just walk into the offices of power and begin to dictate to us.

Finally, we must join forces with other like-minded Americans who are demonstrating a willingness to speak out and to make themselves heard. Not only just to be heard, but also to take control of the outcome of elections through grassroots communication and education about these facts. Currently our most dynamic illustration of such a group is the Tea Party political activist movement who are making America's political enemies tremble.

UNDERSTANDING THE BASICS OF PROGRESS

To be content with the same share year after year is not our nature. There is a basic human heartfelt need to have a sense of progress. This is lacking if there is not an increase each year in our individual share of the world's wealth. With every passing year the earth increases its total number of inhabitants.

Consequently, to satisfy the desires for progress of the present population—plus the new demands that come with each new member to the world population—the need for more wealth increases at a much faster pace than the world population. Indeed, the world at large cries out for more! It is not so much a matter of dividing up the wealth that has already been created. But it is our challenge to create more wealth, at a proportionate rate of increase, to sufficiently satisfy both our individual heartfelt needs and to allow the new larger generations the same opportunity.

This need for progress must not be confused with the term "Progressive." Since the Civil War, Progressives by definition are promoters of Socialism and the Socialist state. Progressives are anti-capitalist and were known at the beginning of the twentieth century to be Socialists. When the term "Socialist" became associated with welfare programs and other state-owned programs to control our lives and our futures, the Socialists changed their name to Progressive. Progressive sounds like someone who supports and promotes the "progress" mentioned above.

The truth is quite the opposite. Progressives support the long-term goal of Communism and state-controlled Socialism and seek to put all forms of the wealth mentioned above into the hands of the state. Progressive Socialists focus their behind-the-scenes attentions on *reinterpreting* our constitution with the intent and commitment to "progressively" expand the power of the central government to accomplish their long-term goal—Communism.

This book contains the truth that destroys the dreams of the elitist progressives who want to control your life and mine. The truth is powerful and strengthens every heart to overcome all obstacles. America is poised to fight back and to overcome the juggernaut of Socialism that has us in its death grip.

We end this book with insights into our opportunities for gaining a newfound political and ideological advantage that we, as American citizens, have never had before, not since before the signing of the Declaration of Independence. However, we must first appeal to your patience and your motivation for seeing your personal and family dreams come true. Not just for you but also for every American who will take personal ownership of their responsibility to be a productive citizen!

Let us begin our journey of learning and equipping ourselves for the battle to come. A good student must be patient and let the learning process blossom into full bloom so that the result is more exciting than they first imagined.

THE CASTE SYSTEM

For centuries the most common answer was a simple matter of caste. If one was born into wealth, then some portion of wealth was guaranteed. Those born to a lower caste could not invade your realm and you in turn were unable to invade that of a higher caste. Thus, the concept of men being "created" unequal financially, through birth, prevailed.

Such a view of mankind, and the subsequent division of the world's wealth in keeping with those views, was held by most for centuries as an undeniable universal truth. Little opportunity was afforded to newcomers on the world scene. One was limited for the most part by the family into which he had been born. Cobblers' sons tended to become cobblers, bankers' sons became bankers, farmers' sons became farmers, and so on. A close-knit social structure was the outcome, very predictable and slow to change. For centuries neither wars nor famines could make a dent in the established order. Then along came new technologies that allowed for a whole different scale of economies. Quality and individuality were overcome by an endless quest for more volume. Thus, the Industrial Revolution was born!

INDUSTRIAL REVOLUTION

At the beginning of the Industrial Revolution, small cottage and personal crafts industries began to be wiped out at a phenomenal rate. The scale of the displacement was such that, within a few decades, the whole of Europe was in upheaval over the social and economic clashes that were produced. We felt little of this in America at first. Now not until the late 1800s did we begin to experience similar unrest and discord. Actually, our earliest pilgrims and settlers were part of the overflow from the English and European industrial development. In America, unlike those old established societies across the Atlantic, we had lots of new land for the taking. The lure of land and a new start far from the effects of the old world centralized industry, with its poor working and living conditions, brought millions of settlers to America.

MANY SPEAK OUT

Throughout the centuries, different voices have been heard lamenting the plight of the working class. However, none had ever been able to effect any major changes in the concept of how wealth is created until the Industrial Revolution. As our own industrial centers began to appear on the East Coast and in the Great Lakes area, labor representatives began to join in a chorus of growing worker unrest, with demands for better working conditions and better pay.

There were many speaking out at that time, but none spoke as eloquently or as convincingly as a German Progressive Socialist politician of the 1840s named Karl Marx. along with his lifelong friend, Friedrich Engels, Marx had spoken out for years against the abuses that were being leveled at the workers of his time. Finally, in 1848, Marx published the results of their joint studies called *The Communist Manifesto*. One of the main concepts of this exhaustive work was their "Labor Theory of Value," which employed the use of many formulas developed by Marx to explain and to attempt to prove the basis for their convictions that all wealth is created by labor. Marx and Engels could find no other solution to the problems of their day than to recommend that all property be expropriated by the state and that the state take direct control of the welfare of the people.

Our postmodern Progressive Socialists, such as our current administration under Barak Obama and their entire ilk, are the intellectual descendants of Marx and Engels. This is true not because we say so, but rather because of their personal admissions. Obama and his advisors have stated as much. They also believe that all property should be expropriated by the state and that the state should take direct control of the welfare of the people.

They do not pretend to do this all at once. On the contrary, their well-studied and executed plans call for the concept of "incrementalism." Incrementalism is the process by which Progressive Socialists have been attacking our beloved constitution for decades and thereby incrementally stealing our rights and our freedoms. They cannot do it all at once—that would show their real intentions and bring about a fight that they could not win. Not until Obama has there been so much transparency from the Progressive Socialists. No, they are generally stealthy and secretive. Only their current major majorities in the House and Senate give them the arrogance to reveal more clearly their true intentions. They prefer deals cut in backrooms and under the table. Transparency and integrity are strangers to them, things that only the non-elite take seriously. Thus, without character or integrity, they seek their ends by any means available to them. In their eyes their ends are superior to such lower caste banalities. So anything and everything is possible for them to do, just so long as it produces the end result that they seek.

THE AMERICAN WORKER

Until the turn of the nineteenth century, American workers were accepted in Europe as just another group of people trying to make a living. No special accord was given them for the growing strength of America's industrial might. Not even America's involvement in World War I was able to impress the Europeans. Yet, without the Europeans fully realizing what was taking place, our post-World War I industrial expansion and development of innovative technologies were setting the foundations of a great industrial power.

The war, which devastated industrial facilities in Europe, inadvertently helped us to expand. In fact, their devastation spurred our expansion to heights which otherwise might not ever have been accomplished in our time. The "American Dream" appeared to be on its way to becoming a reality for all laborers.

DOUBTS ARISE

The world had experienced economic depressions on and off for generations, but nothing as devastating as the one following the crash of '29. When the Great Depression hit, the dreams of millions of Americans were left in shambles. Americans began to doubt our political and economic basis, and many for the first time began to read Socialist and Communist literature. The Bolsheviks had taken over Russia and were talking about "workers' rights" and expropriation of all property from private ownership. It was called a "people's revolution," and was promising to right all of the wrongs created by the Industrial Revolution.

With our own economy in a state of disaster, and so many people out of work (over twenty-five percent at the worst point), many feared that Capitalism had breathed its last breath, and that only a Socialistic form of economy and government could emerge from the rubble.

AMERICA UNDER ATTACK

Our beloved America is now under a full, no-holds-barred, frontal attack by the Progressive Socialists that pervade throughout all political parties and every political office—city, state, and federal—from top to bottom. They pervade the Civil Service system, the Pentagon, and fill the most sensitive positions of every committee on Capitol Hill, in Washington, DC. Progressive Socialists dominate in all major media outlets.

These same Progressive Socialist forces have been at work in the bowels of our university system of academia. They are also in control of our National Education Association, which holds our public schools in a death grip of Socialist ideology. We are blessed only by an outspoken conservative minority in the major parties and by talk radio and cable news outlets such as Fox News. One cannot assume that party affiliation, or even the term "Conservative," is sufficient to know who is pro-Capitalist and pro-America and who is not. Only by understanding what you will learn in this book will you be able to confidently choose between who is who.

THE SOCIALIST PARTY IN AMERICA

By 1924, the Socialist Party in America had reached over one million members. Although feelings ran high in America for Socialist reforms, with more government intervention in the economy being advocated, it was not until the presidency of Franklin Delano Roosevelt in 1932—a Progressive Socialist—that American workers saw the government get deeply involved on their behalf. Still the economy was slow to respond, and never really took off until American factories began to produce weapons for Great Britain in the early years of World War II.

WORLD WAR II

When finally America was drawn into the war on all fronts, every able-bodied man and woman was either in the service or holding down a job somewhere. The resulting benefits were greater than could be imagined! By the war's end, every major manufacturing center in the world had been leveled to the ground except in America. Even in the countries which the Allies had liberated during the course of the war, only the factories that were needed for production of basic war materials had been reconstructed. There was virtually no manufacturing of consumer goods. When the war was finally over, American industry and the American worker reigned supreme over the whole world, and now at last it seemed that nothing could stand in the way of the American Dream for all. Our confidence was at an all-time high and our products were in demand in every nation of the world.

THE RISE AND FALL

With the mighty strength of our industrial and economic base, the lure of Socialist thought declined rapidly. Few—if any—seriously read Communist literature, and our consuming interest was in adjusting to our newfound place of world importance and responsibility. With the world's future peace in the hands of the newly-formed United Nations, the American worker could get on with the task at hand, namely to fill the world demand for American-made products. With the influence of organized labor, workers were setting new standards for wages and benefits. No one could have dreamed such a situation possible only a few years earlier. The American worker had come to believe that no other country in the world could ever challenge the strength of American technology and know-how and, in reality, that impression was not altogether incorrect. However, despite the fact that America still leads the world in technology, we have just experienced another depression surpassed only by the Great Depression, and whole industries are staggering to regain control of their domestic and world markets.

What can we say then about the creation of new wealth? Can it be explained or will it always be out of reach of the average man and woman? Is there hope in our lifetime for every adult person to potentially have access to the power to create wealth, and to satisfy personal desires for progress, while at the same time allowing the newborn to do likewise?

The quick answer is absolutely! The "how to" aspect is not so complicated that you and I cannot understand how to make it work for us. The first thing we know is that government can protect us from foreign enemies and provide important infrastructures. The second thing we know is that government cannot make us wealthy, take away all risk from life, or create one new economically-productive job.

OPPOSING IDEOLOGIES

In order to put these questions in better perspective we need to first look at what general responses there have been to the Industrial Revolution. Before the Industrial Revolution, wealth and its division was more a matter of caste and family than anything else. But with industrialization, two major yet diverse methods of dealing with the wealth thus generated and its distribution among people were adopted.

One method adopted the concept of socialism, collectively owning and dispersing wealth. The other clung to a modification of the earlier concept that emphasized private and individual ownership, with as little disbursement outside of that ownership group as possible. The former developed into what we know today as Communism, while the latter is referred to as Capitalism.

Strangely enough, both of these two major concepts currently distribute the majority of the wealth they create in the form of wages. So, despite the diverse ideologies that separate these two worldviews and economic concepts, their convictions on the creation of wealth and their methods of dispersing it are very similar. So long as wages are considered the most representative form of wealth, there will be little opportunity for alternatives for most people.

This begs the question: why are wages considered by the majority to be the most representative form of wealth and how to distribute it? The American experience is entirely different and has its origins in our nation's colonial experience with personal Capitalism.

Thus, in order to arrive at any intelligent answers for our questions concerning wealth and how to acquire more for everyone, we need to review three of the most revolutionary concepts ever proclaimed about the creation of new wealth and how it should be distributed.

THREE REVOLUTIONARY CONCEPTS

Our first and oldest record of a revolutionary concept of how wealth is acquired is in the Hebrew Old Testament, found in Deuteronomy 8: 17-18:

> You may say to yourself, *'My power* and the *strength of my hands* have produced this wealth for me.' But remember the Lord your God, *for it is he who gives you the ability to produce wealth,* and so confirms his covenant, which he swore to your forefather, as it is today. (Emphasis mine)

The second record of a revolutionary concept of how to get wealth is not too unlike the first. It is found in our American Declaration of Independence, July 4, 1776, where it flatly states that "all men are created equal," as opposed to the assumption of unequal status or class by birth.

The third exception recorded is found in the *Communist Manifesto,* formulated by Karl Marx and Friedrich Engels in the mid-1840s. Their philosophies are generally termed "Communism." Under the discussion of the Labor Theory of Value, Marx says:

> A use-value, or useful article, therefore, has value only because human labour in the abstract has been embodied or materialized in it. How, then, is the magnitude of this value to be measured? Plainly, by the quantity of the value-creating substance, the labour, contained in the article.

(Although Marx is not always easy to read, it will be necessary to quote him from time to time in order to present a clearer understanding of the origin of thought in modern day Communist societies and in the minds of our current Progressive Socialists in the Obama administration.)

Let's consider these three revolutionary concepts of wealth and how to obtain it, beginning with the concepts of Marx and Engels.

MARXIST PHILOSOPHY

Starting with Marxism or Communism we must first address the fact that they are complete philosophies, not just a theory of economics. Marx and Engels believed they had discovered the laws of history and labeled it "Scientific Socialism." They believed they had also discovered the laws of motion and of development of the universe, and they believed they had discovered the laws of animate and inanimate nature.

The basis for their philosophy is "Dialectical Materialism," a phrase meaning that all things can be explained materially, without the presence of any spiritual reality. Such a materialistic concept holds the essential belief that the world of our five senses is all that counts and takes the stand that all happenings in the universe are strictly physical reactions.

Thus, by basing their whole system on pure materialism, Marx and Engels rejected all forms of religion or spiritual reality. For them, the intangible world of the spirit did not exist in any real sense.

They realized, however, that other people believed in God and had various religious orders and customs. But when calculating the value of an article by use of their formula for labor value, their materialistic approach only considers the raw material, plus the market value of the labor time it took to make the article. No credit or attention was given for either personal initiative, or spiritual inspiration with the imagination required to conceive, plan, and develop an idea. It is as if these intangible factors did not exist in the wealth-creating process.

Indeed, all over the world, where this philosophy is promoted as an answer to mankind's economic and sociopolitical problems, religion as a basis for inspiration and direction is put down as a crutch used by the weak and ignorant. Much effort and political power is expended to curtail or eliminate altogether the faith of the people, just as we see in America's system of academia and public education.

UNDERSTANDING OUR CURRENT DANGERS

To fully understand the seriousness of our current dangers and the Progressive Socialist attacks on us and our beloved America we must first understand that all of the Cabinet members—all of the appointed Czars and all of this administration's department heads, including himself—have graduated from the progressive socialist academia system in our highest universities, with honors in these very classes. These individuals are ideologues, totally committed to the Marx's labor theory of value and the so-called "Scientific Socialism." All of the Progressive Socialists from Marx's time to the present have embraced these theories as fact. Where there are differences, they are minuscule, but the end is always the same: total control.

ACT OF ROBBERY

For Marx and Engels, Capitalism as expressed in the Industrial Revolution was fundamentally an act of robbery. Marx lashed out against the rich and the upper middle class with these words:

The bourgeoisie, wherever it has got the upper hand, has put an end to all feudal, patriarchal, idyllic relations. It has pitilessly torn asunder the motley feudal ties that bound man to his 'natural superiors,' and has left no other bond between man and man than naked self-interest, than callous 'cash payment.' It has drowned the most heavenly ecstasies of religious fervor, of chivalrous enthusiasm, of philistine sentimentalism, in the icy water of egotistical calculation. It has resolved personal worth into exchange value, and in place of the numberless- [non-voidable] chartered freedoms, has set up that single, unconscionable freedom—free trade. In one word, for exploitation, veiled by religious and political illusions, it has substituted naked, shameless, direct, brutal exploitation.

Please note these words from the above statement by Marx: *"It has pitilessly torn asunder the motley feudal ties that bound man to his 'natural superiors,' and has left no other bond between man and man than naked self-interest, than callous 'cash payment.'"* *(Emphasis mine)*

What we read Marx saying is that the single unconscionable freedom—free trade—is not really the exploitation (which he uses as a pretext for power), but rather the loss of feudal ties between people like you and me and our "natural superiors." These are strong words, forcefully written and totally believed by Obama and his ilk, who hold themselves up to be our superiors. All the elite, despite their political alliances, hold themselves to be superior to the average American citizen. In their eyes we are ignorant slow types, who need them to care for us and to tell us how to live and how much success we can have.

To be fair to Marx and Engels, we certainly know many modern examples of brutal exploitation in the name of free trade or Capitalism. However, what they witnessed in their day was much more severe, more akin to the atmosphere of concentration camps. But, although they affect our economy and our personal lives, the frailties of human nature and the vices that go with them do not at this point play a part in understanding the principles of the creation of new wealth. Those who understand the true principles of creating wealth not only abhor such historic anomalies, they avoid such. It will be proven that quite the opposite is what assists in the creation of wealth.

LABOR IN THE 1800S

To further understand Marxist philosophy one must examine closely his Labor Theory of Value. This "labor value" as it is called, needs to be examined in perspective, considering the setting in which it was developed.

The mid-1800s presented a bleak outlook for the new laboring class which comprised a major segment of every country in Europe. The average pay was just enough to stay alive. The dignity of the private craftsman was being lost or compromised, for many trades had been rapidly automated and assimilated into large sprawling factory complexes that, in many cases, used more children than adults. The view of the factory owners was that children were agile, didn't take up much space, and could turn out almost as much work as a grown man or woman with less expense.

These factories were surrounded by shanty towns that resembled refugee camps more than communities of families. Compulsory education for children was nonexistent, and women worked fourteen hour days, six and sometimes even seven days a week. Such a large class of laborers was made possible by the relative crudity of the factory design of that day. Early machines were lacking in almost all features of automation or safety and required many simple support tasks that necessitated the presence of a worker.

Trades mechanized the fastest included basic skills such as spinning raw materials like wool or cotton into thread, and weaving it into cloth. Besides these early textile workers, there were many other trades people (cobblers, tailors, smelters, and woodworkers, to mention a few) whose livelihoods were radically transformed. All of these skills had formerly been done by individuals on their own farms or within their own cottage or in small shops, and were all done by hand or with simple hand tools. With the advent of water power, some concentration of work areas occurred and crude machinery began to take the place of the individual craftsman. But, with the advent of steam power, these changes began to accelerate astronomically.

As more advancement was made in machinery that could be run by steam power, new production methods were created. Each of these innovations needed a human being to feed the raw material, to adjust a running apparatus, or to lift a load of finished goods. No special education was needed, and not even the ability to read was essential for the main tasks of production.

Thus, children, women, and men barely existed doing the most menial of tasks, and living in the most meager surroundings with little or no hope of improvement. It was against this background that Marx and Engels formulated their economic and political philosophies.

MEN WITH SERIOUS INTENTIONS

Marx and Engels were not ordinary laborers disgruntled with their lot in life. Marx was the son of a lawyer, and Engels was the owner of textile mills. Regardless of how we might ultimately view them, their conclusions were derived from an intense study of their times and culture, and represented the most revolutionary contemporary political thought of the period.

They worked as a team and Engels actually supported Marx for many of his later years, after his writings and political opinions became so revolutionary that he no longer could derive any income from them. These two men, viewing the situation from intellectual, Progressive Socialist philosophical, and experiential points of view, concluded that the masses had no hope in the hands of the controllers of capital. They concluded that only through a workers revolution could the masses become free.

Through their conclusions, Marx and Engels made the stand that they were the ones who had discovered the source of all wealth. Subsequently they made a most forceful economic revolutionary theory, with applications that greatly impacted their world.

Both men were dead before Lenin was captivated by their revolutionary theory and doctrine. However, unlike how Marx and Engels thought the workers' revolution would start—from the masses of the poor laborers—Lenin took it to the university students and the liberal intellectuals of Russia.

Lenin was then able to forge a subversive movement that captured the newly-won power from the successful Bolshevik revolution against the Czar. He did this by killing the main leaders and assuming the reins of power for himself: but more on Lenin later.

Progressive Socialism—read Communism—has ever since been promoted from within the universities of the world, including our own!

Up until that time the age-old question of "What is new wealth and how does one get one's proper share?" had not been meaningfully addressed in the societies of the old world. In the new world of America and the thirteen colonies, there was a fundamentally different experience, as we will soon illustrate.

Indeed, a contemporary world struggle for control and influence is being spearheaded by this philosophy. One only has to compare the number of emerging third world countries embracing these ideals as their goal and economic standard against those that are accepting democracy and Capitalism with free trade to quickly see to which side the world scales are tipping.

It is because of this world struggle that every American needs to understand the basics of Marxism, Democracy, and Capitalism with free trade, especially as expressed by our founding fathers and protected within our constitution and the Bill of Rights. The Marxist concept is affecting our lives even now, and could conceivably control our lives totally at some future point unless we choose otherwise!

LABOR THEORY OF VALUE

While Marx wrote over 400 pages explaining the Labor Theory of Value, a good understanding of the theory fortunately does not require that much explanation. Simply put, Marx wrote that the value of a given article was made up of two basic elements: one was the raw material used to make it, and the other was the labor expended to fashion the raw material into the article.

At first glance this may seem simple and straightforward enough. More depth is revealed, however, when you consider that all labor is counted, even the labor put into the mining, growing, harvesting, or fashioning of the raw material. The concept of value represented in units of labor—when carried to the logical conclusion from this perspective—excludes the capitalist investor, the inventor, the creative initiative of many different individuals or factory owners and managers entirely. Thus, Marx believed that capital (invested money) was not a factor in the production or creation of new wealth.

Likewise, even the machines employed in the manufacture of goods and services had to be made by someone, and that labor input constituted their total value as well. Therefore, the conclusion was that the "capitalist" (the factory owner and the investors) who did not actually work on the job but only received their income from the capital invested, had in fact stolen income by expropriating the true wealth. This wealth could only justly belong to the laborers.

Consequently, Marx wrote in the *Communist Manifesto* in 1848:

> Along with the constantly diminishing number of the magnates of capital, who usurp and monopolize all advantages . . . grows the mass of misery, oppression, slavery, degradation, exploitation; but with this too grows the revolt of the working class, a class always increasing in numbers, and disciplined, united, organized by the very mechanism of the process of capitalist production itself. The monopoly of capital becomes a fetter upon the mode of production, which has sprung up and flourished

along with, and under it. Centralization of the means of production and socialization of labor at last reach a point where they become incompatible with their capitalist [covering]. This [covering] is burst asunder. The knell of capitalist private property sounds. The expropriators are expropriated.

Are you feeling like one of the expropriated yet?

The one other value added to either the cost or total value of an article was the expense of education and training. Marx and Engels at that early date realized the need for the upgrading of skills and education. We might call that research and development today. Yet they could not imagine the great transformations that such training and research would take!

SOURCES

One might wonder from what sources Marx and Engels drew their ideas. Were they all original? They were not without sources, though some are acknowledged, while others can only be imagined. Interestingly enough, Marx and Engels were not the first to declare that labor was the source of all wealth.

THE MECHANICS' UNION

Twenty years before publication of the *Communist Manifesto,* the *Preamble of the Mechanics' Union of Trade Associations* (Philadelphia, 1827) declared flatly that labor was the source of all wealth. But, instead of demanding the rights to all the wealth that labor produced, they asked only for an equitable share of it, only that which could be (clearly demonstrated to be a fair and full equivalent) for the productive services they rendered. We find this elaborated in the following passage:

> We are prepared to maintain that all who toil have a natural and unalienable right to reap the fruits of their own industry; and that they who by labor (the only source)

are the authors of every comfort, convenience, and luxury are in justice entitled to an equal participation, not only in the meanest and coarsest, but likewise the richest and choicest of them all.

At this point one might ask the questions, "Where did the profit go? Who gets it?" In fact, a profit aspect in the Labor Theory of Value does not exist. Ideally it is all supposed to be divided up equitably among the laborers with the "capital owners" getting nothing if they did not participate in the actual work process. The reasoning thus developed, they concluded that this concept would surely reduce the cost of goods and services, making it possible for the masses to purchase more with their earnings.

This is exactly the position of Obama and his administration. Consider Obama's anger at the bonus programs of many banks, Wall Street firms, and automobile manufactures. Obviously, for Obama and his advisors, profits are evil and the government needs to take them away from the evil entrepreneurs.

DAVID RICARDO

Karl Marx was known to be influenced by David Ricardo (1772-1823), the leading British economist of the early 1800s. Ricardo's Labor Theory of Value similarly held that the value of a commodity is determined by the amount of labor needed in its production. Marx carried this basic idea to its ultimate conclusion in a purely materialistic vacuum.

Unfortunately Marx never had the privilege of seeing the intangible aspects of good management relationships with workers, and later developments of the successful corporate cultures. Without this, and lacking the advantage of accepting the presence and influence of spiritual inspiration and creative initiative as important parts of technological breakthroughs, he was left with few alternatives. From his perspective he was limited to early Capitalism as he knew it (suffering a worldwide collapse) or the state taking over the control and ownership of all private property for the good of the people.

ARISTOTLE

Aristotle is also one of the sources acknowledged by Marx. Living in a time of chattel slavery, where men were used as we use machines today, Aristotle attempted to solve the mystery of wealth and its distribution. His conclusions differ considerably from those of Marx. In fact, after almost one hundred years of Communism in the world, these conclusions compete strongly against Marxism and very much in favor of both Capitalism and free trade. Let's examine these two philosophies more closely to see where they diverge.

EQUIVALENT VALUE

Referring to Aristotle's *Ethics* (Book V, *Justice and Fairness: A Moral Virtue Needing Special Discussion),* Marx discusses the classic question of how we can equate the value of beds and houses so that a certain number of beds can be justly exchanged for a certain number of houses. Marx says that Aristotle recognized that we cannot equate qualitatively different commodities unless we can somehow arrive at a common denominator for them. Marx quotes Aristotle as declaring that "… it is impossible that such unlike things can be commensurable."

Marx then adds that Aristotle himself tells us what barred the way to his further analysis—it was the absence of any concept of value. Aristotle writes,

> What is the equal something, that common substance which admits of the value of beds being expressed by a house? Such a thing, in truth, cannot exist.

Marx delivers what he believes his Labor Theory of Value is a solution to this problem Aristotle (he claims) failed to solve. According to this theory, two qualitatively different things can be made commensurable through the common denominator of labor value, arrived at by measuring the amount of human labor involved in the production of each. Thus measured, things of equivalent value can be justly exchanged.

Perhaps it should be mentioned at this time that Marx was never able to determine the value of a "unit of labor." He expressed confidence, however, that economists would be able to do so. He was wrong: no Communist economist to date has successfully determined the value of a unit of labor. Russian economists worked on it for decades with no further mention on their progress. The issue has now been discreetly dropped.

MARKET DEMAND

When we look further at Aristotle, we find that he does pose a solution. Aristotle does say that a just exchange of qualitatively different things requires that they be of equivalent value and that this in turn requires some way of quantifying their value. To quote the fifth chapter of *Ethics,* Book V:

> All goods must therefore be measured by some one thing . . . this unit *is in truth, demand, which holds all things together;* for if men did not need one another's goods at all, or did not need them equally, there would be either no exchange or not an equal exchange. (emphasis mine)

Aristotle admits, as Marx says, that it is impossible for two different things of different quality to be made "perfectly equal." However he immediately adds what Marx left out: "But, with reference to demand, they may become so sufficiently."

So far as is known, Marx and Aristotle offer the only two recorded philosophical answers to the problem of how to compare the value of two different things having different qualities to determine the unit of measurement for the purpose of justice in their exchange. The question is, which one is correct?

ONLY LABOR

If Marx's Labor Theory of Value is true—if labor alone is the source of all value in economic goods and services—then labor would be entitled, as a matter of strict justice, to all of the wealth produced. Accordingly, labor—in both the living human form and, as Marx claims, in the "congealed labor" (that labor accumulated and congealed in the machines made by labor)—contributes everything to the production of wealth except what nature itself supplies.

If this is true, then Marx could be deemed correct in claiming that "capital property" in private hands should be expropriated. Thus he advocates that the state, having "expropriated the expropriators," should then control all capital instruments for the general welfare of the laboring masses. The wealth produced should then be distributed to them according to their individual needs.

Presumably, any other method of determining values must involve the imposition of an arbitrary opinion of value. Yet, under the labor only value system, all "free marketing" is unconscionable according to Marx. It therefore leaves the determination of what a unit of labor is worth to an arbitrary standard, whether it be the opinion of an individual or a group of individuals.

From these examples we clearly see the absolute centerpiece of the Progressive Socialists: wealth must be distributed according to *need* not according to *ability*. From this comes the phrase: "Not to each according to their abilities but to each according to their needs." In the Russian and Chinese experience it was assumed that the government officials were needier than others, so they got the lion's share!

Can you hear the voice of Obama as he explains to the plumber that "It is good to spread the wealth around." Through the plumber's ability he earned it but Obama wants to steal it from him and give it to those that Obama concludes to have a greater need! Don't be surprised at the arrogance of this. All Progressive Socialists believe that the end by any means is justified: lying, fraud, backroom deals, and stealing, to name but a few.

ONLY MARKET DEMAND

However, Aristotle could also be correct. If, in fact, the free market demand establishes the common ground for comparison of two different things with different qualities so that they can be equitably exchanged, then the more people who hold capital instruments and control the processes of production the better. Such a situation would allow for a much larger number of potential buyers and sellers.

Which idea is correct? Let's continue our examination, and you can determine which one you believe to be correct and the conclusions that you choose to accept.

PRODUCTIVITY OF CAPITALISM

With what we have seen to this point about Marx and Engels' Labor Theory of Value, you could possibly be thinking that they had a very dim view of the power of Capitalism to create new wealth. Such a conclusion would be premature. It is not Capitalism's lack of power to create new wealth that Marx and Engels attack. Rather, it is the lack of equitable distribution of the wealth thus created! Marx shares their view of Capitalism eloquently as he states:

> The bourgeoisie during its rule of scarce one hundred years has created more massive and more colossal productive forces than have all preceding generations together. Subjection of nature's forces to man, machinery, application of chemistry to industry and agriculture, steam navigation, railways, electric telegraphs, clearing of whole continents for cultivation, canalization of rivers, whole populations conjured out of the ground—what earlier century had even a presentiment that such productive forces slumbered in the lap of social labor? *(Communist Manifesto,* 1848)

Not since Marx wrote that tribute has the world heard such public credit for productivity given to Capitalism by any Socialist or Communist leader! Are you listening, Mr. Obama?

COMMUNISM TAKEN SERIOUSLY

At the time the *Communist Manifesto* was written and presented to the world at large, and in particular to nineteenth century Europe and England, Marx and Engels were considered too radical to be taken seriously by the industrial leaders of that time. Marx died in 1883 and Engels survived him another twelve years to die in 1895, neither having seen their philosophies bear any fruit. Not until this philosophy was embodied in the brilliant and forceful leadership of a man called Lenin did the world begin to look seriously at Marxist thought.

Lenin was born on April 22, 1870, in Simbirsk (now Ul'yanovsk), a quiet little town on the Volga River. His real name was Vladimir Ilyich Ulyanov—he adopted the name Lenin in 1901. Historians believe the name may refer to the Lena river of Siberia.

Lenin was a fast learner and began to read when he was five years old. He became involved in revolutionary activities early in his life, and was exiled once from Russia from 1906 to 1908. Returning in April of 1917 to Petrograd, he received a hero's welcome. He then led the Bolshevik party in an unsuccessful attempt to take over the country and had to flee to Finland. He returned again and, in October of 1917, was successful in leading the Bolshevik party's takeover of the country.

The Bolsheviks had come to power with the help of a simple slogan: "Bread, Peace, Land." This slogan had little to do with the theories of Marx, but it had real meaning to starving housewives and their families, soldiers sick of war, and peasants hungry for land. However, true to the Communist philosophy in which he believed, his first official act as the new leader and undisputed head of the Bolshevik revolutionary party, was to abolish private land ownership. Thus, overnight, all privately-owned land became the property of the state.

For the first time the world would get a chance to see the philosophy and economic theories of Marx and Engels put to practice. Lenin was to later describe his own dictatorship as "unrestricted by any laws."

In a later chapter we will discuss and review the record of the Communist experiment and examine what it means to us today.

LABOR AND LEISURE CONTRASTED

In view of the fact that Marx has confirmed a case for labor value, it is important to take a look at some very early views and contrasts between *labor* and *leisure*. For most American workers it is a simple distinction. Labor is something you offer an employer in return for a wage, and leisure is something you do to have fun. Although this is the commonly accepted understanding among the American workforce of today, it is much too superficial, and arises mostly from our limited exposure to the history of leisure in a sociopolitical sense.

The Greek word for leisure, "schole," is synonymous with our English word, "school." It means learning: mental, moral, or spiritual growth. Now, I am aware that in the minds of many red-blooded Americans, sitting back to watch a good football or baseball game on TV is about the most healthy, mental, moral, and spiritual thing one can do! However, contrasted with a viewpoint from the past, Aristotle offers us a challenge to see how unimaginative we are with our so-called "leisure time."

MEN OF LEISURE

Aristotle describes the occupation of virtuous men of property in these words:

> Those who are in a position which places them above toil have stewards who attend to their households while they occupy themselves with philosophy and politics.

The words philosophy and politics are shorthand for Aristotle to describe the Greek view of the activities of leisure—namely, engagement in the liberal arts and sciences and occupation with the institutions and processes of society.

Aristotle further distinguishes between two kinds of wealth achievement when he says:

Accumulation is the end in the one case, but there is a further end in the other. Hence some persons are led to believe that obtaining wealth is the object of household management, and the whole idea of their lives is that they ought either to increase their money, or at any rate not to lose it . . . The origin of this disposition in men is that they are intent upon living only, and not upon living well.

It's amazing how such a viewpoint from the past can so closely mirror our present day dilemma! Much effort today is directed to teaching people that their life's true value lies in more than the simple accumulation of money. Rather, it is the development of the whole person—mentally, physically, spiritually, as well as financially—that is important!

In contrast to such a balanced approach, we see the distinct disconnect there is in the mind of a Progressive Socialist such as President Obama. Obama declared in his speech to the new graduates of Arizona State University that they should not focus on business and financial success. He encouraged them to seek positions in nonprofit organizations and or public service positions in government:

Now, in the face of these challenges, it may be tempting to fall back on the formulas for success that have been pedaled so frequently in recent years. It goes something like this: You're taught to chase after all the usual brass rings; you try to be on this "who's who" list or that top 100 list; you chase after the big money and you figure out how big your corner office is; you worry about whether you have a fancy enough title or a fancy enough car. That's the message that's sent each and every day, or has been in our culture for far too long . . . that through material possessions, through a ruthless competition pursued only on your own behalf . . . that's how you will measure success.

Compare this statement to what Marx said:

The Bourgeoisie, wherever it has got the upper hand, has put an end to all feudal, patriarchal, idyllic relations. It has pitilessly torn asunder the motley feudal ties that bound man to his 'natural superiors,' and has left no other bond between man and man than naked self-interest, than callous 'cash payment.' It has drowned the most heavenly ecstasies of religious fervor, of chivalrous enthusiasm, of philistine sentimentalism, in the icy water of egotistical calculation. It has resolved personal worth into exchange value, and in place of the numberless- [non-voidable] chartered freedoms, has set up that single, unconscionable freedom—free trade. In one word, for exploitation, veiled by religious and political illusions, it has substituted naked, shameless, direct, brutal exploitation.

Obama goes on to say:

> Now, you can take that road...and it may work for some. But at this critical juncture in our nation's history, at this difficult time, let me suggest that such an approach won't get you where you want to go. It displays a poverty of ambition—that in fact, the elevation of appearance over substance, of celebrity over character, of short-term gain over lasting achievement is precisely what your generation needs to help end . . .

> With a degree from this outstanding institution, you have everything you need to get started. You've got no excuses. You have no excuses not to change the world. Did you study business? (Applause.) Go start a company. (Applause.) *Or why not help our struggling nonprofits find better, more effective ways to serve folks in need?* (Applause.) Did you study nursing? (Applause.) Understaffed clinics and hospitals across this country are desperate for your help. Did you study education? (Applause.) Teach in a high-need school where the kids really need you; give a chance to kids who can't get everything they need maybe in their neighborhood, maybe not even in their home we can't afford to give up on. Prepare them to compete for any job anywhere in the world. (Applause.) Did you study engineering? (Applause.) Help us

lead a green revolution (applause) developing new sources of clean energy that will power our economy and preserve our planet.

Please take special note of the highlighted sentence above. It clearly states that in Obama's mind *nonprofits* are a *better* and *more effective* solution for people in need than a successful company or industry that might employ them.

Certainly pure materialism is not the ultimate goal for a truly successful life. But what we have here is a sitting president of the United States, for the first time in our history, discouraging personal financial success while blatantly promoting nonprofit and government alternatives as being better and more effective. Put this in context with the new "jobs bill." How many profitmaking jobs do you really think this bill is intended to create?

Thus, we can summarize by saying that Marx and Engels made an exhaustive argument in favor of labor value as an element in the process of creating new wealth, and that every person has a right to sell or trade that value as a commodity. However, they have left doubts as to whether or not labor is the only value in the creation of wealth. As a result, we refer to labor, physical presence on the job, as only one of *three* elements responsible for the creation of new wealth!

ALL MEN ARE CREATED EQUAL

Our second revolutionary concept: Our Declaration of Independence set forth the most revolutionary concept of basic rights and personal freedoms ever recorded in history. We quote directly from the original manuscript to refresh our memories of these bold and powerful concepts:

In Congress, July 4, 1776. The unanimous Declaration of the thirteen United States of America,

When in the Course of human events, it becomes necessary for one people to dissolve the political bands which have connected them with another, and to assume among the powers of the earth, the separate and equal station to which the Laws of Nature and of Nature's God entitle them, a decent respect to the opinions of mankind requires that they should declare the causes which impel them to the separation,

We hold these truths to be self-evident, *that all men are created equal, that they are endowed by their Creator with certain unalienable rights, that among these are Life, Liberty and the pursuit of Happiness,*

That to secure these rights, Governments are instituted among Men, deriving their just powers from the consent of the governed,

That whenever any Form of Government becomes destructive of these ends, it is the right of the People to alter or to abolish it, and to institute new Government, laying its foundation on such principles and organizing its powers in such form, as to them shall seem most likely to affect their Safety and Happiness. Prudence, indeed, will dictate that governments long established should not be changed for light and transient causes; and accordingly all experience hath shown, that mankind are more disposed to suffer, while evils are sufferable, than to right themselves by abolishing the forms to which they are accustomed. But when a long train of abuses and usurpations, pursuing invariably the same Object evinces a design to reduce them under absolute Despotism, it is their right, it is their duty, to throw off such Government, and to provide new Guards for their future security:

Such has been the patient sufferance of these Colonies; and such is now the necessity which constrains them to alter their former Systems of Government. The history of the present King of Great Britain is a history of repeated injuries and usurpations, all having in direct object the establishment of an absolute Tyranny over these States. To prove this, let Facts be submitted to a candid world.

From this point on, a list of twenty-seven paragraphs follow outlining the unacceptable acts of injustice performed against the Colonies by the king of Great Britain. Thus the drafters of the Declaration of Independence concluded their presentation with these closing words of commitment:

> We, therefore, the Representatives of the United States of America, in General Congress, Assembled, appealing to the Supreme Judge of the world for the rectitude of our intentions, do, in the Name, and by the Authority of the good People of these Colonies, solemnly publish and declare, that these, United Colonies are and of Right ought to be, Free and Independent States; that they are absolved from all Allegiance to the British Crown, and that all political connection between them and the State of Great Britain, is and ought to be totally dissolved; and that as Free and Independent States, they have full Power to levy War, conclude Peace, contract Alliances, establish Commerce, and to do all other Acts and things which Independent States may of right do. And for the support of this Declaration, *with a firm reliance on the protection of divine Providence, we mutually pledge to each other our Lives, our Fortunes and our sacred Honor.*

(The text of the Declaration of Independence given in this quotation follows the spelling and punctuation of the parchment copy, emphasis mine.)

ARISE AMERICA, REBUILD YOUR GOD-GIVEN CAPITALIST FOUNDATIONS

Having weaknesses common to man, several of the signers of the Declaration were owners of chattel slaves. They depended upon this type of "capital instrument" to maintain their lands and to give them the privilege of leisure time to be able to pursue the aspects of politics and civil development. Although you and I could easily agree on the lack of virtue in keeping chattel slaves, I'm sure we could also agree that we are glad that these men were above the need for toil and could pursue such imaginative uses for their leisure time.

In fact, it was the very pursuit of such imaginative causes that led nearly all of those signers of the Declaration who were slave owners to voluntarily free their slaves before the start of the Civil War.

Certainly the architects of America's Declaration of Independence established for all time that all men and women are created equal, and that we each have a right to initiate change in our lives and our surroundings in order to pursue our happiness. In addition, they established the very necessary and important counterpart to the first liberty, namely, the right to engage in leisure activities to ensure the protection of our equality in creation and in the freedom of exercising our *initiative* for the pursuit of happiness. Such freedoms also imply the right to own property and to seek sufficient income to provide for such pursuits. We refer to this element in the process of creating new wealth as our "creative initiative." Thus, we have the second element responsible for the creation of new wealth!

GOD-GIVEN POWER TO OBTAIN WEALTH

For our third revolutionary concept let's consider the oldest recorded claim of how wealth is created. The Old Testament Hebrew record says in Deuteronomy 8:17:

> You may say to yourself, "*My power* and the *strength of mine hands* has produced this wealth for me." But remember the Lord your God, *for it is he that gives you the ability to produce wealth,* and so confirms his covenant which he swore to your forefathers as it is this day. (Emphasis mine)

Assuming for argument's sake that this is a true statement, we would also have to conclude that God was not just talking to the Hebrews to whom this passage is addressed. This logical conclusion is based on the fact that everywhere in the world, no matter where one travels, there is always wealth in the hands of at least a few individuals. They may not respect it or even be aware of the fact that their power to obtain wealth comes from a God who is the giver of all wealth.

Therefore we must also conclude that wealth comes as the result of adherence to some universal law to which God is indirectly referring, a law that is making it possible for them to "obtain wealth" even though they are unaware of any spiritual involvement.

SPIRITUAL STRENGTH

First, let's take note of the position of this individual used as an example in this passage of scripture. This person believes that it was the power of his hand that got him his wealth—that is to say, his muscle power, over which he individually wielded control.

We would do no injustice to the intent of the passage to propose that God has no argument with the individual about the value or importance of his hand or his muscle energy. A point is being made that the driving force behind that hand and muscle was from within the person, in the person's spirit where God could touch him. This passage is telling us that we are more than muscle and bone; there is a creative spiritual force that lies within each one of us. Some have learned to draw on it more than others.

More importantly, however, this passage of scripture is addressing the fact that there is a spiritual part of us, where spiritual forces can work through us beyond our own natural abilities. Now—whether or not one subscribes to the Jewish or Christian faith, or believes in God or a Supreme Being—the reality of man being more than bone and muscle is a truth which few would deny.

It is more than just noteworthy that we are a country which, aside from many human faults, basically believes in God and the power of prayer. More than any other country in the world, America was founded by people who were specifically looking for greater freedom of expression in their faith and religious convictions. This underlying foundation of faith has often been challenged by the pressures and changes we have encountered over the years since the first pilgrim set foot on this continent, yet we keep measuring up to these challenges and overcoming them.

This is especially evident in today's renewal of the search for more reality in life than just money and vocations. Many are finding a great experience in new sensitivities to God's part in the overall make-up of men and women. Furthermore, it is a fact that, when given the resources and the time, no country to date has been able to match America in either speed or quality of new creative technological discoveries in any of the scientific disciplines. This is not because we are the only country with a basic faith in God, since that is not true. Every country on the face of this planet now has some elements of faith in God, and an argument could be made that they are obviously blessed proportionately by the presence of that faith. What I am specifically challenging Americans to address is the reality of our fantastic accomplishments relative to new technology, inasmuch as the whole world has an equal opportunity to perform.

There is a great challenge being presented here to anyone who can accept it, namely that *spiritual inspiration*—and not just for the avowed religious orders, priests, and ministers but for everyone—has a bearing on one's ability to create wealth.

This universal law is open to all, and we have no right to assume that we will not be challenged by other countries. It is no wonder that much of the rest of the world tries, from time to time, to buy or even to steal our technological secrets when they can't discover them on their own. It's not that they would be unable to discover them, given the time and money to do so. It just seems to be a fact of life that we consistently get there first most of the time. Perhaps we should consider our spiritual sensitivity and values more seriously and give thanks for the blessings in the abundance of "spiritual inspiration" we have thus far enjoyed. For example, China has the fastest growing Christian community in the world at this time and it is no secret that this is having a great positive effect on their economy.

To characterize this phenomenon of spiritual intervention into our lives as the third element in the process of creating wealth, we use the term "spiritual inspiration." Thus, we have three elements responsible for the creation of new wealth, and not just one as Marx and others have thought!

SUMMARY

The central core of your life and your liberty is greatly impacted by financial wealth, how it is perceived, understood, created and distributed to you, your children and grandchildren from generation to generation. Neither illiteracy nor ignorance of the truth will shackle you or limit you, as you focus on the value of your Personal Capital Compound. As you grow in your knowledge and understanding of these realities and of your own personal value as a financial instrument of wealth creating value, you will inevitably create an atmosphere of optimism and attract relationships that will open many new doors of opportunity for you.

Capital Compound Theory of Value

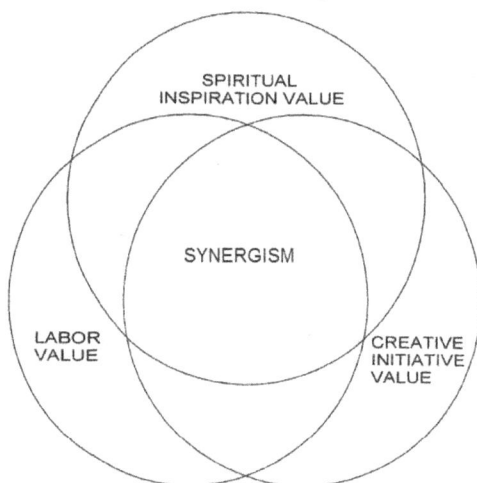

SPIRITUAL INSPIRATION VALUE

SYNERGISM

LABOR VALUE

CREATIVE INITIATIVE VALUE

Chapter Two

Alternatives

When we examine certain realities and we don't like what we see, we are most likely to ask that age old question, "What are my alternatives?" Alternatives are like the "Y" in the road, they give us a choice of directions. The challenge is to pick the correct side of the "Y." One side could be as bad as the current situation we face; the other could or could not take us to where we want to go. How do we chose which alternative is correct and which road is going where we want to go? This chapter will give you the tools you need to understand, not only the quality of the alternatives, but where they will take you!

ALTERNATIVES DO EXIST

We can now address the question of creating new wealth and what possible alternatives might be available to satisfy our needs for a sense of personal progress, while at the same time allowing new generations the same privilege. The present distribution practices of both Communism and what passes for Capitalism offer little chance for the kind of dynamic progress the world demands. The very basis that has been adopted for primary distribution sets up certain expectations and limitations that make it impossible to reach the kinds of consistent growth patterns and equitable distribution methods that are needed to fulfill the demands of our world population. Consequently it should not come as any great surprise that we have conflicts and struggles in almost every quarter of the world.

However a review of our own Declaration of Independence and the Hebrew Old Testament shows that there are elements at work in the process of creating new wealth that were not properly understood or respected by union leaders and men like Ricardo, Engels, and Marx. These elements, Creative Initiative and Spiritual Inspiration (which we will more fully identify) refer to the "intangible elements." These two additionally identified elements are at work whether we recognize them or not, so it makes good sense to attempt to understand them better. Assuming that these elements could be consciously maximized to our personal benefit through such an improved understanding, our alternatives and successes would be altered dramatically for the better.

There are a variety of aspects related to what we have already covered which need to be examined more closely in order for us to fully appreciate what our new alternatives might be.

CREATED EQUAL

Now what about those "self-evident" elements of the Declaration of Independence? First, we'll look a little closer at "all men are created equal." It must be remembered that, at the time of the writing of the Declaration of Independence, the worldview was exactly the opposite—all men were not considered created equal.

Although it might seem to most Americans that the issue is settled now and that world opinion is unanimously in favor of the concept of "created equal," reality demands that we face the fact that much of the world today still suffers from many forms of slavery. You don't have to be in chains to be a slave. Thus, it is important for us to review some of the history behind these conflicting viewpoints.

MAN: A CAPITAL INSTRUMENT

In the days of Aristotle, men harnessed the human workforce with about as much regard for personal rights and personal desires as if they were dealing with cattle. But don't let that mislead you to thinking that either master or slave felt the difference between their stations was permanent or ordained. Not so, for in those days he who was a slave today might be the master tomorrow. His people or original tribe or race could (and frequently did) come and make war against the land and country of his present master, thus liberating the one who was a slave, and in turn enslaving him who only days before was in the position of master. History is full of such examples.

Another misconception peculiar to America about chattel slaves stems from our own limited historic experience with the slavery of the black nations of Africa. When America set up the slave plantation economy of the South, those slaves were almost totally limited to use as physical laborers. Only after many years did slaves useful for administrative and literary work emerge. There was nothing wrong with the slaves' ability to learn or master new skills, but it took time for them to learn our language and culture (without the advantage of formal education) and demonstrate their natural talents.

MORE THAN MUSCLE

This was not so in the ancient and pre-Renaissance days of the Far East, Middle East, and European empires. Slaves were valued as much for their administrative skills (reading, writing, accounting) and their entertainment skills (music, dance, feats of physical prowess) as they were for pure muscular strength. Many military engineers and captured soldiers were made slaves, and then served the new masters as the best source of information about the terrain and geography of the country being invaded.

The Caesars and governors of Rome took delight in exhibiting to one another the best quality of slave possible. They took pride in the fact that they could assign matters of state and finance to such well-trained individuals. Josephus, the talented and famous Jewish historian of the first century A.D., was a captured slave of Caesar. Josephus was given certain powers and freedoms by decree in order to carry out the work of writing for the Roman Empire.

BRUTE FORCE

You might be wondering, in view of the constant potential for sudden changes of status between being slave and master, what in fact did determine the freedom and independence of the individual in those days? The simplest and most direct answer is: brute force.

But that is not the total answer. Every nation had then, even as they do today, two basic goals to consider. First was the protection of the country called home. Secondly, they had to consider the freedoms that were allowed within the borders of that country. Some people could argue that many of the slaves captured by the Roman Empire fared better in the long run as slaves of Romans than they did as citizens in their native countries. I'm not sure that the slave would agree, but if you count only living conditions and the quality of food and material surroundings, you might have a marginal argument.

Thus, if your soldiers were stronger and braver than those of another country and you could conquer that country, you most probably would. War was an accepted method of gaining new territories and new wealth, a fact still not totally eradicated from our present world. The most valuable spoils of war, second only to the gold and silver one could haul away, was the quality of the slaves captured. In short, the slave was a valuable commodity.

The advent of law and the concept that man could govern himself with something more civilized than brute force was the seed that actually grew into our modern methods of protecting our freedom. However, this is true only within national borders. The world family of nations has yet to devise adequate legal structures for the successful protection of international rights and freedoms. The United Nations, our latest attempt to create such a structure, has been marginally successful at best.

MENIAL TASKS

Our present day reliance upon machines to do many of our menial tasks gives us a decided disadvantage in understanding the people of past eras, people like the American white plantationist of the South or the people of high ranking status in the ancient and Middle Ages. What we fail to appreciate is that no one anywhere in our world's history—no matter what race, color, or creed—enjoys or willingly accepts the fate of earning his living by the sweat of his brow or of performing endless menial tasks for the sake of subsistence. In other words, human nature hasn't changed any since Adam and Eve were cast from the Garden of Eden. As the Biblical account goes, they were condemned because of sin to earning their living by the "sweat of their brow," and their descendants have tried to get someone or something else to perform these tasks for them ever since!

MEN DREAMED OF MACHINES

Aristotle made an observation that relates well to the situation in the days of chattel slavery:

> If every instrument could accomplish its own work, obeying or anticipating the will of others... if the shuttle could weave and the plectum touch the lyre without a hand to guide them, chief workmen would not want servants, nor masters slaves . . . instruments are of various sorts; some are living, others lifeless; in the rudder, the pilot of a ship has a lifeless [instrument], in the lookout man, a living instrument; for in the arts the servant is a kind of instrument. . .

> (An economic) possession is an instrument for maintaining life. And so, in the arrangement of the family, a slave is a living possession and property a number of such instruments; and the servant is himself an instrument which takes precedence over all other instruments.

Aristotle—a great and highly respected man in his day, and far removed from the mechanized and computerized world in which we live—could nonetheless relate to a future possibility when tools could move under their own power and intellect, and would not need slaves to give them power and life.

A DRASTIC ALTERNATIVE

In such times it also was not uncommon for a man to actually "sell himself" as a servant to another. If one found himself without the means to provide properly for himself and his opportunities appeared bleak, then such an alternative was not unthinkable. Remember, the servant or slave was of great value as a possession. Therefore, if the only asset remaining to you in such a society and such times was your worth as a slave, you might have to consider the option of selling yourself to meet your obligations.

Different countries handled that situation in different ways. Some of the most detailed laws governing the purchase, treatment, and sale of slaves are recorded in the Old Testament. We will look at one passage recorded in Exodus 21:1-6 that clearly indicates that the slave in question was willingly offering himself for sale.

> Now these are the laws you are to set before them. If you buy a Hebrew servant, he is to serve you for six years. But in the seventh year, he shall go free. If he comes alone, he is to go free alone; but if he has a wife when he comes, she is to go with him. If his master gives him a wife and she bears him sons or daughters, the woman and her children shall belong to her master, and only the man shall go free. But if the servant declares, I love my master and my wife and children and do not want to go free then his master must take him before the judges. He shall take him to the door or the doorpost and pierce his ear with an awl. Then he will be his servant for life.

There were many other laws that governed the handling of servants and slaves. What this clearly shows, however, is that there could easily have been worse fates than being a servant or slave for life.

A CAPITAL INSTRUMENT

Our reason for examining such factors is not to justify chattel slavery but rather to emphasize the value of human labor and its consideration as a capital instrument in times past when there were no machines to employ in the service of others. Slaves took away the drudgery of menial tasks and served the function of profit-making for their owners. Though we are "free" men and women today the value, of our Labor, our Creative Initiative and our Spiritual Inspiration have not changed. Nor are we any less a capital instrument today than at that time. This is a very important point which we will discuss further in later chapters.

Despite our "enlightened" society and the absence of slaves in modern industrialized Western countries, it is not impossible to consider that our societies of today could return to some form of forced dominion over less-fortunate people for the purposes of using them in chattel servitude. I believe the only preconditions necessary to bring the world to such a state would be the loss of our artificial power sources.

If all such power sources should ever be lost (electricity, hydro, internal combustion, solar, nuclear, and so on) and the world's power was reduced to that of only animals and humans, you could rely on the return of chattel slavery. In fact there are many refugees that tell of conditions resembling slavery in many of the third world countries that have an overabundance of people and a scarcity of alternative power.

The point is that all of history teaches us that human labor was and still is considered *a* commodity of value to be owned, contracted for in exchange for money and/or benefits, and at times even sold or traded. This is certainly the definition of a capital instrument even as a mechanized tool is a capital instrument.

NEW CONTRIBUTION

After much to do about their so-called scientific discovery of the Labor Theory of Value, Marx and Engel's only real new contribution was to argue forcefully for the right of every worker to own his or her own labor value, and thus, to be able to trade it as a commodity in the production process. As for being the first to actually recognize such a basic right, they were far behind the Hebrew law, among others. They did, however, serve to inspire the workers of their day and many millions since then in an effort to squeeze more wages out of the owners of industry for services rendered. But these laborist wages do not reflect a full appreciation of the power of the individual to create new wealth. Factories can be and are being closed, workers are being laid off, even complete occupations can disappear—where is the wealth and security in that arrangement?

NEW SOURCES OF ENERGY

History shows that, even before the time of Marx and Engels, certain creative minds and imaginative leaders—such as our own colonial architects of the Declaration of Independence, Bill of Rights, and our Constitution—were recognizing that man is more than just an "instrument for labor." They were beginning to perceive the power of something intangible about mankind, things like his creative initiative and spiritual inspiration. The evidences of new power sources and mechanical advantages—such as the advanced technologies of the water wheel, the metal plow, and many other basic tools that so dramatically changed their world—inspired the declaration that "all men are created equal," and sounded the death knell to the class society.

Even with today's mind boggling evidences of technological breakthroughs due to creative initiative and spiritual inspiration, we still have a majority of people fooled by the myth that "labor" is the only source of wealth or at least the primary source. Truly, the founding fathers of our country were inspired men.

UNALIENABLE RIGHTS

Now we can examine the other self-evident truth, namely "that they are endowed by their Creator with certain unalienable Rights, that among these are Life, Liberty and the pursuit of Happiness."

An endowed right to life, liberty, and the pursuit of happiness given by one's Creator gives the individual an independent right to own and sell or trade his labor value as a commodity in the production process. But more than labor is implied here. The pursuit of happiness is going to require that one be diligent in the use of not only his muscles, which produce pure "labor value," but also in the use of creative initiative and spiritual inspiration, which directs the muscles to perform the same tasks more efficiently through change and invention. Implied too is involvement in the "leisure activities" of society and the perfection of public and private institutions to help preserve and improve one's opportunity to create new wealth.

Our founding fathers have another truly revolutionary concept to offer. It is in the genius of recognizing the unalienable right that one possesses beyond the mere marketing of his labor and intangible elements. The founding fathers recognized, for the first time, the importance of one's right to protect his use and free interchange of those values and the property they would produce!

LABOR VALUE EXPROPRIATED

In sharp contrast to the sensitive insights on individual rights expressed in America's Declaration of Independence, Bill of Rights, and Constitution, under the Marxist concept of labor value, one only has a single opportunity to cast his vote. That is when he raises his voice in favor of the government body, to which he will surrender his labor value and everything else. Remember that in the Marxist philosophy of labor value, all private property is to be expropriated from the capitalist. And, according to Marx, the labor value of the worker is his only commodity, and it too is a part of the capital instruments of industry. Proof of this is in the fact that when the state expropriated all capital instruments of production and all private property, they did not fail to expropriate the labor and creative value of the worker.

At the same time, they once and for all closed all his doors of opportunity to change the bargain by political process. In contrast we are still blessed to have that option.

THE WHOLE IS GREATER THAN
THE SUM OF ITS PARTS

In an effort to get a new perspective, let's examine these three revolutionary concepts—labor value, creative initiative, and spiritual inspiration—as a compound instead of limiting our perspective of them as always being separate. The idea of bringing these three elements together in one compound should not be too difficult because each of these elements is an integral part of everyone. As human beings, we are the organic and spiritual embodiment of these three elements. Each and every person in the world can and does express these three elements, albeit in varying degrees and intensities.

A compound can also demonstrate a very special characteristic that we express as individuals to some degree, and as groups of individuals to an even greater extent. This characteristic or phenomenon is called synergism. Synergism is present when two or more basic elements are combined to create a new compound with a weight greater than the sum of all of the individual elements' weights. The difference between these two measurements is referred to as the "synergistic effect." Something extra is created out of combining these individual elements that cannot be obtained by using the elements separately.

Therefore, when an individual undertakes a given task there is much more at work than just labor or muscle. All of the opportunities for implementing one's creative initiative and spiritual inspiration are at work, and something more takes place—synergism. Sometimes it is only a simple task, and one doesn't have to demonstrate much effort or power from this compound in order to get the desired results. At other times the task may be very complex, requiring maximum effort and power output from this compound to get the desired result. If there is a failure to accomplish the goal, one should not consider that they personally are a failure, but that their compound experience requires more time and intensity before it will measure up to that particular task. In fact, sometimes a task can only be accomplished when tackled by more than one individual. This joining of forces allows for extra synergism and we get a Joint Capital Compound.

We believe that the evidence overwhelmingly supports the basis for the creation of new wealth through this compound, and not just as a single element of labor value (which over half of the world believes and practices as part of a philosophy called Communism, and which we in America are now facing in the onslaught of Progressive Socialists). America now believes as the Communist and Socialist world believes, and we practice laborism in the quest for full employment. Sadly, this quest is used as a smokescreen to promote government-centralized power in the name of labor contracts, bank bailouts, and industry acquisitions. Everything is done to emphasize wages and benefits as the primary method of distributing the wealth of a given company or entity.

THE NEW COMPOUND

What does our new compound look like, now that we have joined all three elements together?

1. Labor as a value, as a capital instrument and as a commodity of exchange or sale.

2. Creative Initiative as an adjunct to labor and the unalienable right to use my creativity in leisure activity, to protect my ownership and the control of both my labor value and my creative initiative.

3. Spiritual Inspiration (faith and resources in gifts) on which to draw for the power and inspiration of my creative initiative.

CAPITAL COMPOUND THEORY OF VALUE

What shall we name this? Because it is made up of more than one element it is a compound, and because it is a commodity of each individual and can be traded or contracted for hire we shall refer to it as the Personal Capital Compound of each individual.

1. Its first element is our *labor value*. This is basic Capitalism in a solid state. Any physical action on our part is part of the labor process. This is the visible and tangible element which heretofore has been incorrectly accredited with being the sole source of wealth. In order to express any new approach or to experiment with several variables we will use our labor element. We have direct control over this element of time, place, and practicality. We can decide from time to time how much we want to utilize it.

2. Its second element is our *creative initiative value*, protected by our equality and unalienable rights. This is basic Capitalism in its liquid or more pliable state. We have the opportunity to apply our creativity each and every time we are confronted with a task. We can choose to follow the established procedure usually applied in the accomplishment of any given task in question or we can decide to think of a better and more efficient way to accomplish the same task with greater results. Creative initiative takes us outside of the box we are not locked into a predetermined limitation.

3. Its third element is the *spiritual inspiration value*. This is basic Capitalism in the gaseous, intangible state. This is the least understood element, and the demonstration of its state does not appear to be in our direct control. Nevertheless, each and every time we are faced with tasks that we recognize to be beyond us, we tend to seek for outside inspiration and strength. Sometimes

it comes to us quickly and forcefully and at other times seems to be elusive and distant. Despite our difficulty in sensing its presence, the Capital Compound would not be the same without it. Through this powerful element we have been gifted by God with our nation's Capitalist Foundations.

SYNERGISM

When there is a compounding of the three elements into the Personal Capital Compound, synergism is created. This is greatly increased in the Joint Capital Compound presented below.

JOINT CAPITAL COMPOUND THEORY OF VALUE

The Personal Capital Compound expresses the compounded value of the first three basic elements for creating all wealth. This applies to each and every individual in the world. However, no man is an island or lives alone in the process of life. Therefore, when two or more individuals work together they form what we call the Joint Capital Compound. When there is a compounding of two or more Personal Capital Compounds into one focused effort, sole proprietorship, corporation, organization, or institution, the synergism is *exponential* compared to just the personal capital compound. Later we will use this term when we speak of operations involving more than one individual.

PROFIT AND ADDED VALUE

Taking the position that this Capital Compound is a more correct model of the source of all wealth than any which limits our focus to just one of its elements (such as labor), we ask the question, what will it do for us? When viewed as a compound, a living organism, and not as a single element, Capitalism gives us the key to the mystery of new wealth. Instead of the limited productivity of just a single element, labor, we realize that there are really three elements at work.

There is also the additional power and value of the synergism that our three elements produce as a byproduct of their joint activity. This synergism of our Capital Compound gives us more than just the sum of its parts: we now have true added value which explains how we can constantly increase our production to meet our individual desires for progress and, jointly, those of others and of our nation.

For example, should a given product into which we've invested our Capital Compound have a market demand that offers us a price above the cost of our total expenditure of time and energy and all related costs, we can call the difference our profit. If for some reason there should be no demand for our product thus formed by our Capital Compound, we would still be able to recognize the fruit of our synergism and its effects of increasing value, but we would not be able to declare a profit. In fact we might have to declare a loss.

In this case there is either no market for our product/service or we have been inefficient in our efforts, meaning that someone else can produce the same or similar products or services and do so at less cost. So we see one of the basic advantages of a free trade environment. The competition that manages to use their Capital Compound to produce products or services more efficiently, coupled with an equal efficiency in marketing (another use of the Capital Compound), will set the standard for all new wealth to be created in that market area.

NEW WEALTH DISCOVERED

Now that we have discovered the secret of the creation of new wealth, a very pointed question comes to mind: How much of this wealth have you included in your contract bargaining? Are you being compensated for all aspects of your Capital Compound contribution? Or are you like so many others, receiving compensation for only one element of your total Capital Compound, the labor value element?

Those few in our world who are being compensated for all aspects of the three elements of their Capital Compounds are the true Capitalists of the world. They not only have discovered the secret to the creation of new wealth, but they hold within their power important concepts, insights, and economic instruments for equitable distribution. These are the keys to the future destiny of every American!

THE HAVES AND THE HAVE-NOTS

In a world of haves and have-nots, the haves are clearly those who receive compensation for all three elements of their Capital Compound, and not just for their labor alone. In sharp contrast to these few are the more than 150 million wage earners of America. As we know it today, wealth and financial security for the American worker and for the workers of the world relates to the single element of labor: one's physical presence on the job. This single form of compensation provides for both their present needs and their future hopes and dreams. It seems inconceivable that such a limited distribution of the total wealth produced by the Capital Compound should provide the only hope of the world's workers to someday be considered a have.

Sadly, less than five percent of Americans are directly involved in any entrepreneurial endeavor through which they derive the majority of their income. Compare this to the fact that, from the founding of the first colonies to the signing of the Declaration of Independence, over ninety-five percent of all Americans were directly involved in entrepreneurial enterprises through which they received their income. In short our forefathers, the great pioneers of America, were small business owners and therefore, true Capitalists.

The enormous productivity power of the three elements of the Capital Compound—Labor, Creative Initiative, and Spiritual Inspiration— are our tools of liberation from state-controlled economies dreamed of by the Progressive Socialists. America must rid herself from the control of these elitists or suffer the consequences of being duped by the pied piper of Socialism. America knows the way: we have the blueprint and the examples of our forefathers. We have the vision of our own hopes and dreams but as yet we have not realized the substance of those hopes and dreams!

SUMMARY

We have chosen a good and well traveled road. This is a road that has blessed all travelers of every generation with personal freedom and respect for others. It produces limited government, equal protection and justice under the law. It produces personal and joint financial strength, beyond imagination

THE THREE ELEMENTS OF WEALTH

1. Manual Labor

2. Creative Initiative

    We hold these truths to be self-evident, that all men are created equal, that they are endowed by their Creator with certain unalienable rights, which among these are Life, Liberty and the pursuit of Happiness.

1. Spiritual Inspiration

    You may say to yourself, my power and the strength of mine hands has produced this wealth for me. *But remember the Lord your God: for it is he that gives you the ability to produce wealth*, and so confirms his covenant which he swore to your forefathers as it is this day. - Deuteronomy 8:17

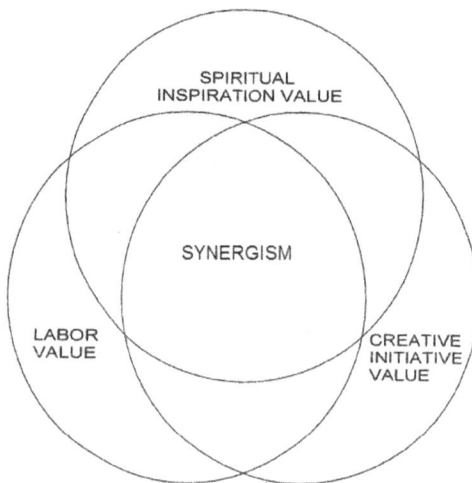

Capital Compound Theory of Value

Chapter Three

Controlling Wealth

We all remember the oft-used expression "follow the money." Controlling wealth is at the heart and core of the human struggle to gain wealth. Man is infected with an insatiable desire to control his life through financial means. Man's desire to control his financial wealth is an overwhelming force that requires predetermined structures, wisely designed, to ensure that we do not lose that which we have rightly earned, and to protect us from both ourselves and others against the forces of destructive greed and obsessive control.

THE DISCOVERY

Discovering the secret of the creation of new wealth is terrific! It's like uncovering a rich vein of gold. Knowing that it exists and that riches are a possibility conjures up exciting visions. One thinks of all manner of things to do with such power, both selfish and noble. Everyone has, at one time or another, dreamed of what they would do if they suddenly came into possession of a fortune.

THE DREAM

Throughout the centuries, most of mankind has only experienced wealth as a dream. In fact, even the discovery of the secret of new wealth is of little value if all one can do is look at the gold discovered, but can't touch it or use it. What is missing is the ability to control and utilize our newly-discovered secret.

THE QUEST FOR CONTROL

Looking backward from our new perspective, we find that from the earliest beginnings of mankind, use of the Personal Capital Compound and the quest to control it are evident. All races and all tribes have left records of these efforts. Throughout many stages of increasing sophistication—from hunter, agrarian, craftsman, trader, and conqueror, to the most modern, state-of-the-art businessperson— man has staked out his Personal Capital Compound territory and defended it. We call it our "niche in life" or our "occupation" but it is really just our Personal Capital Compound manifesting itself!

OWNERSHIP

The animal kingdom uses everything from camouflage to brute force in protecting both individual and territorial resources. Although resembling the animal kingdom at times, in his use of force, man has developed a system of ownership as his method of control. In the earliest records of even the most primitive societies, ownership in some form is evident. Cuneiform tablets dating back thousands of years record the deeds and titles to lands and houses. The earliest known inscriptions are from the lower Tigris-Euphrates Valley in Mesopotamia, now Iraq. Tablets found in Uruk (Warka) and nearby sites may have been written as early as 3,000 B.C.

Ancient Hebrew laws of the Old Testament record the importance of honoring the boundaries of private property:

> Do not move your neighbor's boundary stone set up by your predecessors in the inheritance you receive in the land the Lord your God is giving you to possess. – Deuteronomy 19:14

LAND AS A CAPITAL INSTRUMENT

Beyond basic personal possessions—such as clothing, cooking utensils, shelter, and weapons—land was the first major capital instrument to be privately owned. It was land that afforded man his first real opportunity to utilize the Personal Capital Compound to consistently create new wealth. In the *Report on Manufacturers* issued by the Secretary of the Treasury in 1791, Alexander Hamilton summarized one of the explanations for the greater productiveness of agricultural labor, maintaining "that in the production of soil, nature co-operates with man; and that the effect of their joint labor must be greater than that of the labor of man alone."

THE INDIVIDUAL—MORE THAN MUSCLE

If considered purely as an instrument of muscle power to perform work, without consideration of the power of his creative initiative to increase the efficiency of the task and without access to spiritual strength and inspiration, man is one of the weaker sources of such power. However, given the opportunity to fully utilize all elements of the Capital Compound, man can use the soil of the earth to develop very large amounts of wealth.

In Chapter One, we brought out that the unalienable rights stated in the Declaration of Independence imply responsibility for the use of our labor, creativity initiative, and spiritual inspiration for involvement in the leisure activities of society for the perfection of public and private institutions. One of the refinements of public institutions is a system of ownership. Such recognition of ownership allows control, and control gives rise to the possibility of security, and the hope of security encourages involvement. Consequently, it is reasonable for us to be more protective and diligent with that which we consider to belong to us as opposed to that which belongs to someone else.

Ownership and similar institutions of society constitute the basis for the various types of civilizations that exist around the world. It is important for our study of control to consider some of the basic elements found in all cultures, from the most primitive to the most advanced.

No other society has been more attentive to the protection of private property and its use by the owner than America. Likewise, no other society, having reached such extraordinary levels of private property success as the American experiment, has suffered such major reversals as we now suffer under the onslaught of the Progressive Socialists.

CULTURES AND SOCIETIES

The concept of "culture" as presently held today originated fairly recently. The first definition of culture as a sociological term was that of E. B. Tyler, a British anthropologist. Tyler wrote in 1871:

> Culture . . . is that complex whole which includes knowledge, belief, art, morals, custom, and any and all other capabilities and habits acquired by man as a member of society.

Tyler considered culture as the sum total of human achievements. Additionally, during the 1920s the Functional Anthropologists, headed by A. R. Radcliffe-Brown and Bronislaw Malinowski of Great Britain, emphasized the fact that each culture is an integrated whole, like a living being. This latter fact will become more important as we develop our study of wealth and its equitable distribution.

ELEMENTS OF CULTURES

All cultures have certain things in common. For example, anthropologists have discovered that four main elements are found in all cultures: (1) technology, (2) institutions, (3) language, and (4) the arts.

Technology, as anthropologists use the term, includes what can be called the basic equipment for everyday living and the way people utilize it. It has been discovered that all human cultures include the use of tools. Scientists in search of the origin of man have never found a group of people who lacked tools, even if only sticks and fire. Fire is the one tool common to all cultures. Technology can be as simple as a sharpened rock or as complex as computers.

Institutions are groups of people who either live together or meet regularly for some purpose such as establishing or maintaining laws, community responsibilities, trade, war readiness, or religion. A family structure and some form of religion are two institutions common to cultures. In all free societies men, women, and children live together in families.

Language is a third element that all cultures have in common. Each language provides more than just a method of communication. Flexibility in a language makes it easier for a culture to adjust to changes in the environment and its needs. A broad vocabulary allows a culture to express complex ideas plainly. It is interesting to note that, compared to many languages, English has a relatively simple grammatical structure which gives it both the capacity for a large vocabulary and high flexibility.

The arts exist in every culture, and include myths and tales of ancient heroes, dancing, acting out legends and rituals, paintings, sculpture, singing, instrumental music, and various combinations of these forms of expression. Like language, art and music serve as a form of communication, one not constrained by rules of grammar.

CUMULATIVE EFFECT

Technology, institutions, language, and the arts: these four basic elements of every culture are produced as the result of groups of individuals using their Joint Capital Compound and working together. Therefore the product of a family, culture, or society can best be expressed as having been produced by their Joint Capital Compound. Since they are largely interactive in nature, it is impossible for one person to develop these four basic elements alone. We may personally learn and grow from our own experience in these areas, but they can obviously never become elements of a culture or society without the willing participation of others. We see then that the Personal Capital

Compound has a cumulative effect when we, as individuals, function as partners within cultures and societies.

COMMON SENSE

The deep impressions of any cultural cumulative effect are exemplified in one's perspective on common sense. I personally experienced this when I worked in Mexico for about twelve years as a young man. When I first arrived, I thought that I carried with me a certain degree of common sense. I'm sure that all of us have used the expression "Oh, that's just common sense" but I wonder how often we have stopped to ask ourselves what we mean by "common?" To whom is it common?

Actually, when I found myself in the midst of such a different culture from the one I was raised in, I finally had to face that question squarely! My conclusion led to a breakthrough for me in my relationship with the Mexican people: I came to realize that what I thought was common sense was only so for me and those of my culture. Some of the things I did seemed pretty stupid to those residents of my newly-adopted culture. I also learned that some of what I at first thought was backward behavior was really very practical in their environment.

ACQUIRED SENSE

Fortunately, some of the things that I had carried with me into the Mexican culture were needed by them, things I had acquired as a result of living in my own culture. Many things that I knew and at first attributed to that nebulous virtue of common sense I later came to realize were things I had actually *learned* from my parents and associations which they in turn had learned from their parents. To my amazement, I learned that so-called common sense isn't common at all—it's an *acquired* sense through the cumulative effect!

Each bit of knowledge that we hold dear as common sense may not be very significant in and of itself. Yet, taken as a whole, these little cultural nuances make up the unique characteristics and total fiber of any given culture or society. We know that several different subcultures can and often do exist within the same society. American society certainly has many different subcultures within it: each subculture is separate, but bound together with the others in a unique family sharing of these countless common sense understandings.

Oh that President Obama and his administration could demonstrate to us their acquired common sense. I do not personally, nor do I know anyone in any country where I have ever lived or traveled, consider it common sense to spend money that they don't have to gain wealth that they never earned and cannot identify.

Obama and his ilk—*read Progressive Socialists*—steal our personal credit in the form of debt instruments and levy taxes which we cannot pay, to finance jobs that they cannot identify. This is very simple for the Progressive Socialists. In the mind of a Progressive Socialist like Obama, his thinking processes are not based on common sense as you and I have acquired—they are based on his blind faith in the Communist ideology. This Communist ideology believes that government is better equipped and better prepared to run industry than are the private individuals who own it, i.e. stockholders and managers. One glaring illustration of this fact is the absolute disconnect evidenced in Obama and his administration in their strenuous efforts, using every means available, to pass a National Reform Health Care Bill that over sixty percent of the country has emphatically rejected. Their stated reason for this was that they knew that the country wanted it and needed it. This is political and personal arrogance of epic proportions.

They steal our credit in the form of debt and taxes, such as the $850 billion dollar stimulus package and use it to buy controlling interest in industries that they do not know how to run. We must not lose sight of the fact that in the mind of all Progressive Socialists, this is common sense. We must understand how their brains function in order to decipher their lies and their efforts to manipulate our lives. We cannot stop them if we do not understand them for who they are.

EFFICIENCY EQUALS SOPHISTICATION

Remember these four basic elements of society: technology, institutions, language, and the arts? These four basic elements of every culture are produced as the result of groups of individuals using their Joint Capital Compound and working together.

These four basic elements therefore become parts of the Joint Capital Compound and subsequently become the instruments of that culture and society. The relative efficiency with which each individual uses the Capital Compound, and the cumulative effect of that individual efficiency within the culture or society, determines the general level of sophistication of that culture and society.

Thus, cultures and societies which have not developed their Joint Capital Compound very efficiently might be termed backward. Others who have developed their Joint Capital Compound to very high levels of efficiency we call advanced. The real battle, however, is won or lost on the level of the individual. When key individuals within a society are Progressive Socialists, it is impossible for them to win the battle over economics, i.e. the creation of wealth and its equitable distribution, or to lead us into victory with any of their ideas. All of their promises and ideas are based on a false foundation of centralized power and punitive actions against all who would exercise their own Personal Capital Compound for personal profits.

INSTRUMENTS OF THE CAPITAL COMPOUND

All elements of any culture or society, regardless of whether we only view the four basic ones or if dozens can be identified, are the creation and product of that culture or society. Therefore, all the tools that any society has ever developed or discovered—except for labor, creative initiative, and spiritual inspiration, the three elements of the Personal Capital Compound—are instruments of the Joint Capital Compound. This list is long. For example: all institutions (social, political or religious, public or private); gold, silver, precious gems; all natural resources, tools, machines, and all developed technology are instruments of the Personal and Joint Capital Compounds.

It may be truly said that, through control and application of the Capital Compound Theory of Value, man has dominion over all things. These all spring from the productive efforts of any given society's Joint Capital Compound. Consequently, as instruments they must always be subservient to mankind in general, and to individuals in particular. Each individual person is the embodiment of his own Personal Capital Compound or collectively with others as a Joint Capital Compound. We must always remember we are more than instruments and that our personal dignity and our very being should never feel threatened by the physical and creative expression of our Personal Capital Compound. It is honorable and desirable that every individual be successful in the management and use of their own private Personal Capital Compound.

In contrast, Progressive Socialists will always attempt to impugn our personal success in managing our own Personal Capital Compound. They ridicule such success as selfish and wasteful. They use such tactics to justify their punitive taxes and their doctrines of redistribution of wealth. Can't you still hear the voice of Senator Obama, running for President when he said, "It's good to spread the wealth around!"

Is this what the nation thought he meant by "Change that we can live with"? Did the nation understand the mind of Obama and his Progressive Socialist followers, who consider it good to take away the utility and profit of our own Personal Capital Compound and squander it on those who refuse to successfully manage their own? Undoubtedly, the Progressive Socialists find it very good to spread the wealth, so long as it ends up in their bank accounts and those of their friends.

TECHNOLOGY

Let's look closer at the basic element of technology. Technology in its simplest form is the development and use of any tool other than the hand. However, considering this aspect more deeply in light of our knowledge of the importance of creative initiative and imagination in the workplace, we can say technology really begins in the mind. The most basic tool, even a sharpened rock, is impossible to create without first visualizing that tool in the mind. When any tool is implemented, we are utilizing more than just muscle power. Today we have tools that make tools automatically with advanced computer technology—but all existed first as only elements of creative thought. Many different people chose to exercise their creative initiative and spiritual inspiration in concert with the other elements of their Joint Capital Compound in order to bring these tools into existence.

MACHINES

Tools become more advanced as they are developed to do specialized tasks. The more highly developed mechanical tools are called machines. Simple machines include the lever, the wheel and axle, the pulley, the inclined plane, the wedge, and the screw. No matter how complex, all modern machines are based on these six types of simple machines.

MECHANICAL ADVANTAGE

The ability of a machine to do work is measured by two factors. These factors are known as *efficiency* and *mechanical advantage*. The efficiency of a machine is determined by the ratio of the energy it supplies to the energy which must be put into it. A simple lever is a good example of a machine that has a high efficiency. The work it puts out is almost equal to the energy it receives, with very little loss, but no machine is one hundred percent efficient because friction uses up some part of the energy input. However, the energy that is put into a machine gives back a greater reward through what is called the *mechanical advantage*.

For a simple illustration of mechanical advantage, let's use our lever. Let's say it is a crowbar and we will use it to lift four hundred pounds with only one hundred pounds of pressure applied. We simply place one end of the crowbar under our four hundred pound object and place a support, called the fulcrum, one fourth the length of the crowbar from our four hundred pound object. We can now lift the object by applying a pressure of one hundred pounds to the extended end of the crowbar.

By virtue of many individuals exercising their Joint Capital Compound to bring about these great breakthroughs in basic machines we can say that the "mechanical advantage" is an example of measurable synergism.

THE CAPITAL COMPOUND PRODUCES LEVERAGE

The foregoing observations should help us to appreciate the usefulness of the Capital Compound in its most basic form. Yet, no matter how basic its use it always has the same three elements: muscle (or energy), creative initiative, and spiritual inspiration. It is easy to see that in the simple state of a small agricultural community of ancient times, control of technology was in the hands of the individual. Even the craftsman who had given up the land entirely was in possession of his Capital Compound and had individual control. In theory, such a balanced situation seems ideal. However, we are leaving out some very important factors. We must balance this ideal situation with the reality of the other three basics of culture and society.

LEISURE ACTIVITIES

Institutions, language, and the arts are a part of the whole, and cannot be overlooked in the equation. In order for societies to develop beyond basic cultures, some additional activities have to take place besides working the land, developing better tools, and accumulating more goods. Yet we all know that, if matters were left to some, the accumulation of goods, things and possessions, would be the only result of their group effort. As Aristotle said, "The origin of this disposition in men is that they are intent upon living only, and not upon living well." However, history has taught us that no culture or society is possible without the four basics of technology, institutions, language, and the arts.

Thus man found from the very beginning a need for leisure activity that is the effort which must be put forth into the fabric of society other than for the production of material wealth. This included such important matters as protection from danger (the reality of war and of being threatened by others has always been with us). Community leadership in the rules of conduct, spiritual leadership providing moral guidance, and insight in the development of art, philosophy, and politics is always needed.

ACCEPTING RESPONSIBILITY

Who accepts responsibility for this role of the leisure activity for the good of the community? Again we refer to Aristotle's words: "Those who are in a position which places them above toil have stewards who attend to their households while they occupy themselves with philosophy and politics."

Take special note of the dramatic comparison here between *toil* and *philosophy and politics.* Now we can appreciate the ancient concept of the words "leisure activity" as active, time-consuming, responsible activity. Working the land and doing endless physical chores from dawn to dusk was not easy, and man has never been satisfied with limiting himself to physical activity alone. Thus, to the Greeks, leisure simply meant working with the mind rather than with the hands.

THE ADVENT OF THE PROFESSIONAL POLITICIAN

The actual selection process used to determine who was going to be involved in the leisure activities was simple—the person or persons who controlled the largest amount of Joint Capital Compound simply assumed the responsibility. Thus, from the earliest times to the present, with few exceptions the possession and control of a large portion of a community's total Joint Capital Compound has implied the right to participation in the leisure activities. As societies developed into ever larger, more complex systems, those controlling large Joint Capital Compounds began to hire others to represent them in the increasingly demanding affairs of the state. Thus we see the emergence of the professional politician.

A man one evening was watching a national TV news program while sitting next to his wife on the couch. A group of politicians were listening to a speech by a member of the administration when one of them yelled out "Liar!"

The man looked at his wife and said, "Honey it always confuses me when someone yells 'liar' in a group of politicians. I can never figure out which one they are talking about."

Our founding fathers did not think of politicians as "professional" but rather as citizens genuinely concerned and committed to a temporary service of their country. We need to revisit this original understanding.

SELECTION BY FORCE

A question that arises naturally from this model is: how did some acquire control over enough Joint Capital Compound to be able to elect themselves to serve in such leisure activities, and hence to assume a role of leadership within the group?

Evidence strongly indicates that in the earliest cultures, brute force through physical superiority was the commonly employed method of getting ahead. Perhaps the leader was the best hunter, the strongest man in the group, or a combination of strength plus intelligence. In these cases the group benefited from his protection. Perhaps the male mystique of "machismo" has its roots in such early tribal arrangements. The cliché that "might makes right" also seems to illustrate this train of thought.

Our own present-day world conditions provide evidence that perhaps mankind still depends more on threat of force than on negotiation and statesmanship! Our current administration under Obama's leadership and those of his Progressive Socialist team is one very good example in point. As we just said, with few exceptions, the possession and control of a large portion of a community's total Joint Capital Compound has implied the right to participation in the leisure activities. Thus we see how the Obama administration has assumed the "right" to dictate how they are going to serve us, whether we like it or not. What they don't already have in their control of the

Joint Capital Compound they are busily acquiring by taking over industries and claiming their superiority as elitists to amass ever more of America's Joint Capital Compound.

It is critical to understand that no centralized power or government can manage a majority of any nation's Joint Capital Compound without exercising massive controls over every aspect of each citizen's personal life. We, you and I, are these citizens that they seek to control. It is equally critical to understand that Obama and his team of Progressive Socialists cannot take away control of our Personal Capital Compound, and thus our Joint Capital Compound, unless we refuse to take control of the democratic process by which we have the right to regain control of these Capital Compounds and change once and forever the direction of America. At this time in our history, Obama and all of the Progressive Socialists in all of the institutions of America, *do not* have the means to keep us from voting them out of office and thus out of power!

Our future, the control of our own Personal Capital Compound, the Joint Capital Compound of our society, and our ability to lead America into the greatest future ever, is within our hands. Only defeatists, those who have lost all hope of recovering our beloved America, can imagine themselves to be without power and opportunity. These alone could allow such reprehensible and miserable excuses for leadership to destroy our personal lives and the great heritage of our wonderful America!

UNEQUAL ABILITIES

There is another less physical reason why a greater accumulation of wealth was realized by some than by others, and this has to do with the simple fact that not all people have the same abilities relative to the same tasks. Such has been and always will be the case. Therefore, those who were able to most efficiently utilize their total Capital Compound resources through the balanced use of all three elements of labor, creative initiative, and spiritual inspiration, gained control of more wealth more quickly. This would have been true even if each family unit had never needed the help or cooperation of any of the others. However, since abilities do differ widely among individuals, the talents and strengths of one help to offset the weaknesses of others and vice versa. Thus, a community has developed that has certain interdependencies and yet it functions as a single organism.

UNEQUAL DEMAND

The fact of differing abilities leads those individuals with a high degree of managerial and leadership talent to accept the services of those who do not want or who do not feel comfortable in such roles. This relationship then leads to the more complex problem of what is fair compensation for one who does not accept the role of leadership and what is fair for one who does.

It's Aristotle's beds and houses all over again. If both functions were counted as equal, we would simply divide the wealth equally. But recall the words of Aristotle in Chapter One: "All goods must therefore be measured by some one thing" and "this unit is in truth *demand* which holds all things together; for if men did not need one another's goods at all or did not need them equally, there would be either no exchange or not an equal exchange."

The commodity of leadership or management is always, by definition, less plentiful than the commodity of muscle, even though it is very creatively applied. As a consequence, by virtue of demand, the person with the most leadership and management skills will always be successful in getting a larger share of the wealth. The importance of this principle should not be underestimated.

EXPROPRIATION

The problem that has not been properly addressed in these examples is the fact that no one, no matter how menial the task, contributes *only* pure muscle or labor. Each and every person brings with them some portion of the total Joint Capital Compound. In addition, the abilities of the individual that goes beyond muscle and for which no proper compensation has been made will remain with the project's (eternal quality) to be a blessing to its owners and or managers, and a loss to the one serving them. Such acts are indeed an act of expropriation.

This injustice is magnified greatly when we consider the fact that the end result of the action of the Capital Compound is subject to a synergistic effect. This effect produces far more than any one of the elements, or persons involved, could do alone. For this reason, many companies pay bonuses to those who play a significant role in the development of new technologies or formulas. This form of inequity can usually be handled through contracts. When this is not possible, the individual who feels the pressure of such exploitation is free to disengage from that entity and create a new entity which they can then control.

SYNERGISM

The reality of this effect reminds me of the farmer who was very diligent in the management of his property. His farm was very well tended, with beautiful pastures, healthy animals, productive orchards, straight rows of blossoming fruit trees, and white fences gleaming in the sunshine! A stranger happened to pass by one clear day, and seeing such a beautiful sight was moved to stop the farmer in his work next to the fence and congratulate him.

The stranger said, "Sir, God certainly has blessed you with a most wonderful and beautiful farm."

The farmer paused, took off his hat and wiped the sweat from his brow. "Yes," he replied, "God certainly has blessed me and my family, and this farm, but you should have seen it when he had it all by himself!"

This synergism and its contributed value to the total project—in this case, working the land—is left for the benefit of the owner of the land. Such accumulation over time will serve as a growing asset for the original owner and create a concentration of wealth worth far more than the value of the laborer's original input.

THE COMMUNIST EXPERIMENT

If we assume Marx's Labor Theory of Value to be the correct measurement of value in this case we will give the owner compensation for only the actual physical labor that he has put into the project. We could not allow compensation for the other two elements of the Capital Compound because, according to Marx and Engels, labor is the *only* element. Marx would demand much more—in fact—he would advocate that the landowner be paid nothing, and that the workers of the land simply expropriate the land from the landowner.

When Lenin came along and actually carried this philosophy beyond theory and into action, what did he do? The slogan of the Bolsheviks, "Bread, Peace, Land," appealed especially to the starving housewife and the peasant who wanted land. Land, that one great capital instrument, has helped to make men independent from earliest times.

So what did Lenin give them? Well, the housewives continued in near starvation for many years, and food continued to be in short supply for decades in many parts of Russia. As for the peasant, he got no land. In fact all of the land was expropriated by the state as Lenin's first official act after taking power in October of 1917.

Perhaps it should be mentioned at this time that, according to official government reports during the reign of the Russian Communist government, the farmer in Russia was allowed to use only three percent of his allotted land for his own personal production. With the production from this three percent he was free to market it anyway he pleased. This afforded himself and his family the opportunity to maximize their family's Joint Capital Compound with just that three percent of land. It is a powerful testimony to the reality of our Capital Compound Theory of Value model that the Russian farmers during that time produced thirty to thirty-five percent of the whole nation's entire farm production, all from just the three percent of the land utilized as private land. Yet, with all of the power of the Communist state behind them, pushing as hard as it could, they could not (or would not) produce more than just sixty-five to seventy percent from the remaining ninety-seven percent of the land controlled by the state!

TEACHERS (RELATIONSHIPS HAVE CONSEQUENCES)

It is a sobering thing to realize that Obama and his entire administration, including all of his special Czar appointments, have all studied at the feet of Lenin, Stalin, and Mao through studies in our own university academic system! They applaud them and look up to them with reverence. Is it any wonder that Obama and his administration act like they came from another planet? We ask ourselves, from where did they come up with that idea or that policy? Without understanding who their teachers were and whom did they consider to be their most valuable friends and relationships, we cannot understand it.

TOTAL EXPROPRIATION

As we pointed out in Chapter One, Lenin's Socialist state expropriated the labor commodity of the peasant and also of everyone else in the country. Actually, we can now carry that one step further as we have a better understanding of the total Capital Compound Theory of value. One could say that the state expropriated everyone's labor, creative initiative, and spiritual inspiration—lock, stock, and barrel!

In addition to the expropriation of the individual's Personal Capital Compound, they also expropriated the cumulative effect and synergism of the individual's compound and of the Joint Capital Compound of society. This is evidenced by the fact that, once the Communist party took over the leadership of the country, they also abolished those "unalienable rights of life, liberty, and the pursuit of happiness" that our Declaration of Independence says are "self-evident." It is these rights that allow us the privilege to pursue our "leisure activities" of politics and the betterment of the public and private institutions in order to protect our own Capital Compound from expropriation.

One of the central important tenets established by our Declaration of Independence and of our Constitution is the clarification to the whole world that we do *not* receive these rights from *government*. We have them by the grace of God as unalienable rights, *not* by the will of any government!

It is also these rights that we are now in the process of losing through the expropriation tactics of a large army of Progressive Socialists who permeate every nook and cranny of our society. Overall, this "army" is not as large as it is influential. But what it does control is very powerful and important. It controls large segments of our media, academia, local and state governments, our foreign policy centers, our state department and our federal government on all levels.

THE STATE BECOMES THE ONLY CAPITALIST

In the Communist experiment we see that a Communist philosophy leads to "central control" and "central ownership." Yet it's plain that within any society or economic system the persons who benefit most from the flow of the Joint Capital Compound are those who are in control. If we allow the state to control the total Joint Capital Compound of the people and all capital instruments, the state then becomes the only capitalist. Hence the people of Russia or of any other country (our own included) where such an equivalent system exists become merely laboring servants of the state. Some might prefer to call them slaves.

As astounding as it might seem for many, America is now traveling this road to serfdom. All that is needed is for the sleeping eye to awaken and see and for the sleeping mind to awaken and think. It is only a well-informed, focused, and committed electorate—which means you and me—that by the grace of God can and will pull us back from the brink. Arise America and rebuild your God-given capitalist foundations!

SERVANTS AND SLAVES

The subject of slaves and slavery is basic to understanding what we need to know about control of our Capital Compound and capital instruments. The Capital Compound as the embodiment of all wealth becomes the focal point of control but if we wish to control the Capital Compound completely it would be necessary to have exclusive rights to all of its elements. In other words, to own it.

Each individual person represents all three elements of their Personal Capital Compound by virtue of the fact that they control their own labor, creative initiative, and spiritual inspiration. To obtain exclusive rights to the use of these individually-owned capital assets we would have to approve of chattel slavery, the owning of another person. The third value of the Capital Compound (spiritual inspiration) we cannot "own" in the same way that we could the person but by virtue of controlling the person in whom that element resides and is working, we would benefit in the same way as if we could actually say we owned it.

A good illustration of this is recorded in Genesis 30:25 - 30. Jacob, one of the Jewish patriarchs, had given himself in willing servitude to a Syrian named Laban. His agreement had been to work for him for fourteen years in payment for his two daughters, Leah and Rachel, who were to be his wives. At the end of the fourteen years Jacob asks for permission to leave with his wives. This is the record of that event:

> After Rachel gave birth to Joseph, Jacob said to Laban, "Send me on my way so I can go back to my own homeland. Give me my wives and children, for whom I have served you, and I will be on my way. You know how much work I've done for you."

> But Laban said to him, "If I have found favor in your eyes, please stay. I have learned by divination that the Lord has blessed me because of you." He added, "Name your wages, and I will pay them."

> Jacob said to him, "You know how I have worked for you and how your livestock has fared under my care. The little you had before I came has increased greatly, and the Lord has blessed you wherever I have been. *But now, when may I do something for my own household?"* (Emphasis mine)

We see clearly in this case how the fruit of the servant's Capital Compound accrues to the Master. For this reason, Jacob appeals to Laban to give him his independence so that he could concentrate on building his own wealth for his family.

SLAVERY ACCEPTED AS NECESSARY

The basic attitude of the ancients, who were very familiar with slavery, was to accept it as not only necessary, but right. In the environment of the ancients, as cultures possessed the available land and acquired increasingly larger territories, more and more "slave instruments" became necessary. Working the land was difficult enough for a man and his family, but there was much that needed to be done besides working the land. Tools were simple, and although they made the landowner more productive, they could not make it possible for him to produce all that was needed in a growing society.

Because of their apparent differences in natural talent, many men were willing to work as hired servants for those who showed promise in management and leadership. But the compensation for this servitude was small by comparison to the amount of the contribution. The person who willingly gave up his portion of land because of either a lack of desire to be totally responsible for it, or because of a genuine interest in another direction (such as hunter, warrior or craftsman) often paid a great price in security. The old saying was "No matter what happens or what fails, in the end there is always the land." When one worked for someone else there was always the possibility that in bad times one might be asked to leave. A man in such a case, if he had no land of his own, was instantly reduced to begging or charity if he could not find work elsewhere. There were no unemployment benefits, short or extended!

Thus, as communities became larger and needed more managers and leaders to help structure their political, social, and religious institutions, they quickly rationalized that any lack in the workforce could and should be made up in any way possible. Debtors who borrowed from others and could not meet their obligations could willingly give themselves into servitude to their creditors, hopefully to be freed after the debt was paid. Frequently they were simply taken into custody and sold to satisfy their debts. Prisoners taken in battle, if not killed, were sometimes offered as slaves and sold to the landowners.

INSTRUMENT OF LUXURY

One can well imagine the benefit of owning slaves in such times. For a fee one could obtain a valuable instrument of work and production that could be maintained for the price of food and shelter alone, and it even came equipped with "automatic features." Slaves could think and remember instructions to perform at a later time. In contrast to our twenty-first century lifestyle with automated home appliances, home computers, computerized offices, and automated production lines, the ancient's only alternative to doing it all himself was to accept slavery. Crude and repulsive as it might seem to us, for them it was the first attempt at "robots with brains," or your own "home computer" to figure out all of your household needs and carry them out. What's more it could even reproduce itself if you would provide a mate!

"How dehumanizing," I hear someone say. Absolutely! Slavery is just that! Any kind of slavery is an act of dehumanization. A part of that person's "life, liberty and pursuit of happiness" has been claimed as the property of another with no just compensation being made. Such a vital portion of a person's life, once taken, can never be returned even if the person is no longer treated as such. Time has passed never to be returned. We will see in a later chapter how that time is one of the instruments of the Capital Compound and how it participates in the creation of wealth.

America is being assaulted on all sides by Progressive Socialists who see no injustice or lack of common sense in the expropriation of our labor value, our inalienable rights, our creative initiative, and our spiritual inspiration. They have put their hooks deep within our bowels. These self-appointed masters have taken upon themselves far more than the delegated power our republic with its system of checks and balances ever intended to afford them. With sweet talk and smooth arguments they lead us ever closer to the brink of serfdom as surely as if we lived in a time of the ancients.

Arise America, rebuild your God-given capitalist foundations!

LACK OF INSIGHT

Aristotle would have needed a crystal ball to be able to foresee the massive accumulation of creative initiative and spiritual inspiration that was to be visited upon the earth from his time to ours. Although Marx and Engels were in a position to view the tip of the iceberg, they were nevertheless no better able to imagine the outcome.

The root of their problem was a lack of insight into the reality of the Capital Compound, especially its intangible aspects. It was there for them to see even as it is today. It is rarely witnessed in its most productive form, such as the colonists of the first 150 years of the American experience lived it.

This is primarily due to two basic human weaknesses: greed and pride. Of our many human frailties, these two are at the root of our lack of appreciation for the beauty, simplicity, and power of the Capital Compound when fully utilized.

"OURS" IS STRONGER THAN "MINE" OR "YOURS"

You see, when I am more concerned about "me" and "mine" than I am about the collective "ours," I lose all perspective on the one product of the Capital Compound Theory of Value that is most important. I lose sight of the synergistic effect produced by working together, which makes our efforts worth more than the sum of our individual Personal Capital Compounds. No really dynamic creation of wealth is possible without it!

The Apostle Paul, in the New Testament letter to Timothy had this to say:

> People who want to get rich fall into temptation and a trap and into many foolish and harmful desires that plunge men into ruin and destruction. For the love of money is a root of all kinds of evil. Some people, eager for money,

have wandered from the faith and pierced themselves with many griefs. - I Timothy 6:9-10

Paul is not against money or material riches. He used both in his lifetime. What he is warning against is the lust or greed that drives people to think in terms of "me" and "mine." Remember it is the ego that says I and by *"my power and the might of mine hand have gotten me this wealth"* and leaves out completely the reality of the other two elements of the Capital Compound.

This is not to say or suggest that Aristotle, Marx, Engels, or many others before and after them who did not perceive the power of the Capital Compound, were any more or less greedy than others of their day.

What needs to be emphasized is that one finds, both then and now, in the majority of cultures and societies, an overemphasis on "me" and "mine," rather than on "ours." Some might be tempted to say, "Yes, but why not improve on that by first emphasizing the second person, yours, before you emphasize the third person, ours?" A good question that has a direct answer: "Yours" is simply the other side of the coin "mine."

There is certainly synergism evident in the single unit or individual expression of the Personal Capital Compound. But that alone, whether "yours" or "mine" is weak, just as one drop of water by itself is weak. However, when many drops—or many individual Personal Capital Compound building blocks—are put together, the effect is to increase the synergism exponentially. It explodes with creative energy and force!

Think of the effect of only one person working on the international space station program. How much progress could they make alone? Then remember the last time you watched on television as the space shuttle roared off the launch pad, and all of the synergism that is represented. Imagine the billions of details that have been attended to by literally hundreds of thousands of people working together to make it happen. The greater value that is expressed in that aspect of ours overshadows anything that one could possibly gain by insisting on just mine or yours!

Taken in the sense of creating wealth, think of how much income you would earn if you received ten percent of what one Personal Capital Compound could create, versus ten percent of what ten Personal Capital Compounds working as a Joint Capital Compound could produce. Which experience will give you the most return for your time and effort?

A WORLD CHALLENGE

Although mankind is and always will be in some state of imperfection in this life, it is not to assume that improvements cannot or should not be made. Such a narrow, pessimistic viewpoint is exactly what kept Marx and Engels from seeing the future impact of the Industrial Revolution. They were of the opinion that it was only a matter of time before the evils of the ever-increasing power of the mass accumulation of capital and capital instruments in the hands of a few magnates would drive the whole capitalistic system into worldwide bankruptcy.

This is the exact same position of President Obama and his administration. According to Obama and his advisors, the Bush administration had allowed the major corporations, oil companies, banks, auto manufacturers, insurance companies and all health care providers to accumulate so much capital and power, that they were on the brink of worldwide bankruptcy. This concept is right out of the Communist Party playbook. Unlike President Obama, Marx and Engels predicted that at that point the proletariat (the common people) would simply rise up and take over. Lenin in essence was saying with his revolutionary actions, "Why wait for world bankruptcy? Let's expropriate all property and power for the state now." This feeling of Capitalism as some kind of evil element in the world (with all its so-called dark trappings of private ownership and private property) is the very nucleus of much of our present world unrest.

The Obama administration has basically done the same thing. They have nationalized the auto industry and through the bailout money attempted to expropriate the banking industry. However, unlike the prediction of Obama and his advisors, the banks were not on the brink of bankruptcy as proven by the fact that they returned all of the money, plus interest, in a matter of months. A couple of banks have taken longer, but by and large, when the banking leaders saw that the true objective of the administration was to expropriate them through strict controls, they took swift action to give the money back. If they had been close to collapse as they said they were and as the country was told that they were, it would have been impossible for them to return such huge amounts of money in a matter of only a few months.

In view of our present world where diverse opinions and feelings which are stimulating wars and revolution all around the globe, what can we say to this challenge? Is Capitalism evil and communism pure or vice versa? You decide. Capitalism in its complete form as a Joint Capital Compound has a long history of overcoming the tendency of human weaknesses. It supports private property. It supports the individual rights to life, liberty, and the pursuit of happiness. It gives them the opportunity to rule themselves. Communism has a long history of supporting expropriation of property. It denies individual rights to life, liberty, and the pursuit of happiness. It has the world's highest record for the imprisonment and enslaving of countless millions of people around the world.

Communism or Progressive Socialism is a socioeconomic mutation that attempts to make mankind pure in and of himself, without God. This is institutional self-righteousness at its worst and, of course, fails. It never worked in the past, it does not work now, nor will it ever work. You must make your own determination and take action accordingly.

The historic record is irrefutable: the Communist experiment, while claiming to have overcome the inherent evils of Capitalism, has in fact embraced as the only source of wealth the most limited, callous, and inhumane aspect of the Capital Compound—pure labor—and its resultant materialistic expression. Our creative initiative and the blessing of our spiritual inspiration are priceless, both to the individual and to society as a whole.

CONTROL AND SELF-DETERMINATION

Basically it all boils down to a matter of control. When a government, a large corporation, or any central power controls all aspects of the Capital Compound they become the master capitalist and we the people become their servants or slaves. We lose our enthusiasm for life and being productive. We are not encouraged to produce as much materially, nor can we. We are stifled in our desire for self-determination, which is an expression of self control. No one enjoys the feeling of being victimized by some arbitrary power.

Even though such pure materialism ignores two-thirds of the Capital Compound when applied on a broad scale to a whole society, it cannot help but achieve some measure of productivity. However, the results are minimal because of the restrictive nature of such an environment. The example of the Russian farmer is an ideal illustration of this paradox.

FREE TRADE AND HUMAN WEAKNESS

Marx and Engels believed that the gross ills and evils which they perceived in their day were the basic product of that one "unconscionable freedom—free trade," which we quoted from Marx in Chapter One. According to Marx and Engels, this had created the scepter of an ever-increasing concentration of power and capital into fewer and fewer hands. In their eyes such a concentration brought out the worst and most evil aspects of the men in power. In reality however, free trade was not to blame. The true culprits were greed and pride, spawned by the great profit opportunities of the Industrial Revolution. The lust to be rich at someone else's expense gave rise to some truly inhuman abuses of both power and wealth. Few exceptions to this were to be found in the 1800s, but a germ of honesty and integrity, one more powerful than Marx and Engels could foresee, survived that and led to better things with a more efficient use of the Joint Capital Compound.

The modern corporation, with rules favoring the voice of stockholders in annual board meetings, has assisted in the control of greed. However, much more important is the reality that all of the major corporations, with all of their good and bad idiosyncrasies, only produce thirty percent of our GNP. The extraordinary reality is that the Capital Compound Theory of Value is venerated in the success of the small business owners. Collectively the small business owners of America produce seventy percent of our Gross National Product and provide the source of over ninety percent of all new jobs created. Hello Mr. President.

Contrast this with the "wisdom" of the Obama administration as they throw money—trillions of dollars—into the creation of more and bigger government and call it "job creation." There are only two kinds of jobs that any government can create: one is a soldier, the other is a bureaucrat at some level. They may be a civil servant, politician, public school teacher, state university teacher, or law enforcement officer but all are dependent on the federal government. All are dependent upon taxes as the source of their income, whether at the city, county, state or federal level.

Understand this: the government can create "jobs" that do not produce profits. As we just said, they can increase the number of our troops in the military, which is usually good. They can increase the number of our national security forces which is usually good. They can increase the number of our civil service members which is usually not good! They can increase the number of politicians and administration staff to push more and more papers, as they justify their importance in the ever-expanding government which is definitely not good!

Whether for good or not, the government cannot now, nor ever will create a new job that actually engages in a successful *for profit* activity. If we are to enjoy our inalienable right to life, liberty, and the pursuit of happiness now and for our children's future, we must embrace this limitation of government as an absolute principle. We do not have the luxury of holding this truth as merely a highly-held opinion: this principle is so important that we must defend it at all cost, even to the giving of our life to defend it, such as our founding fathers did!

This very principle was at the center of the commitment and dedication of all of our founding fathers. Recall Patrick Henry, who said simply: "Give me liberty or give me death."

ALTERNATIVES

Control and ownership of our own Personal Capital Compound is one of the most important factors we can have to avoid an ultimate total central control. We also need the willingness to work together for the common good, utilizing our "leisure activities" for the development of strong relationships within our corporations, unions, trade circles, and our institutions, to effect the necessary changes

SUMMARY

WE THE PEOPLE

We have "Followed the money." We have seen how that controlling wealth is at the heart and core of the human struggle to gain wealth. Man is infected with an insatiable desire to control the financial means of his life. This is a good thing, but it is such a powerful urge and stressful concern that wisdom must be exercised, to successfully acquire this reality for anyone who will. It is natural for men and women to desire to control their financial wealth. Yet, the wisdom we need to accomplish this with appropriate financial structures and safeguards is in the development and application of predetermined financial structures! When properly designed they insure that we do not lose that which we have rightly earned and to protect us from both ourselves and others, against the forces of destructive greed and obsessive control! In later chapters we cover these in more specific detail.

There are positive solutions to the problems and the dangers we face. It is up to us—WE THE PEOPLE—to determine that we will learn what we have to learn, embrace the challenge to make changes with courage and realize that together, by God's grace, we have our "labor," our "creative initiative," and our "spiritual inspiration" and we will overcome!

Chapter Four

Dangerous Assumptions

Remember the Donner Party? They were the group of pioneer settlers on their way to California and they were taking the trail that led through what we now know as "The Donner Pass" just west of Truckee California. The Donner group estimated the trip would take four months to cross the plains, deserts, mountain ranges and rivers in their quest for California. Many other families joined them and the entire group numbered several new wagons, with eighty seven people in all. Their assumptions were found to be dangerously wrong. They had also assumed that there would be ample hunting game for providing the large group with food. This too proved to be dangerously wrong. From October 28, 1846 to March of 1847, forty one persons died and 46 survived. The survivors escaped death by cannibalizing those who had died. Tragically, all of these wrong assumptions were made on bad advice in order to save time. Progressive Socialists always say that no matter how bad their assumptions are we would all be worse off if we had not done it their way. The easy road of letting someone else carry the burden of our responsibilities always leads to disastrous conclusions, such as the Donner Party suffered.

THE ONLY ANSWER

The solution that Marx and Engels proposed as "the only answer," the total expropriation of all capital and capital instruments, relied heavily on one very dangerous assumption, that the ultimate and total concentration of capital into the hands of a few magnates of industry *was evil,* while the same total concentration of capital and capital instruments into the hands of the state *was not evil!* Obviously, Marx and Engels trusted their philosophy and their ideals of the leisure activities of bureaucrats more than they did the integrity and management skills of the industrial magnates of their day.

This has Obama written all over it! Obama and the elites that surround him, Progressive Socialists all, are deeply committed to the ideology of socialism. By definition, this demands the concentration of mass amounts of financial capital instruments, the country's Joint Capital Compound and the personal capital assets represented in the lives of as many people as possible. Being the wonderful, extremely blessed, and even "anointed" elites that they are, it is only natural and makes perfect common sense for them to control everything! Right?

CHANGES WERE SLOW

In the early years of man's experience on this planet, technology lumbered along at a snail's pace. Each new discovery was slow to be accepted and implemented, more probably because of resistance to change than because a new invention lacked merit. Centuries passed before any really fast-moving changes began. Until the time of the water wheel and the steam engine, only the use of sail and animal power had been available to replace the muscle of man.

As the world population slowly experienced a net increase (real growth after war, famine, and disease had taken their toll) the emerging cultures and societies were able to produce enough additional goods and services to meet the growing demand without undue stress. In addition to meeting the minimum requirements for subsistence, a small net gain was consistently made that increased the total wealth. Those cultures and societies that were successful in utilizing a greater percentage of their Joint Capital Compound were the ones who realized the greatest net gain.

PRODUCTION EXPLODES

When the breakthrough finally came and the first mechanical source of power—steam—was discovered, things began to happen almost overnight. This fast-moving portion of the Industrial Revolution was soon to be followed by the discovery of electricity, and later the internal combustion engine. The emergence of this new applied technology created a chain reaction that caused the Joint Capital Compound to literally explode onto the production scene!

There is an inconvenient truth which the Progressive Socialists do not want you to know about. The areas of the world that experienced the greatest growth through this new era of technology and the applied Joint Capital Compound were in the societies that also embraced the Christian faith. It is no accident that our beloved America—established on Christian principles of faith, integrity, and a strong work ethic—has outproduced every other country in the world.

MARX AND ENGELS REACT

When Marx and Engels perceived that this historical phenomenon was in progress they were horrified. They saw the logical end as a mindless and abandoned use of mechanical power to disrupt what had been a lazily moving cultural system—albeit with little opportunity for growth, but steady and dependable.

Remember these words from Chapter One?

> The bourgeoisie, wherever it has got the upper hand, has put an end to all feudal, patriarchal, idyllic relations. It has pitilessly torn asunder the motley feudal ties that bound man to his "natural superiors" and has left no other bond between man and man than naked self-interest, than callous cash payment.

Natural Superiors in our day should be understood o mean Progressive Socialists.

THE EFFECT OF PERSONAL WEALTH
AND RELATIONSHIPS

We need to examine this historical distrust in the dynamics of the Joint Capital Compound economy and its many players from two perspectives.

"Feudal ties that bound man to his natural superiors" was more than an employee/employer relationship. In most cases the relationship was more paternal, especially in those societies that were accustomed to maximizing their Joint Capital Compound, inasmuch as they knew how. There existed a bond of trust. The ideal tradesman or farmer was the one who took care of his people through thick and thin. This is a type of "job security" created by those close paternal ties. Now all of this would be upset. The new rulebook hadn't been written yet, and the prospect was every man for himself, or so thought Marx.

Marx believed that countless craftsmen were soon to be put out of business and all would be reduced to "cash payment." In short, it meant a total reordering of personal relationships in exchange for the new high technology of that period. The simple answer to the problem was for government to expropriate everything and take on the paternal role of Big Brother. Marx and Engels saw no inconsistency in this answer, nor could they foresee the evils that would be produced by such arrogance and lack of faith.

Based on all that we now know about Obama and his advisors, they don't see any inconsistency in this massive takeover anymore than Marx did. Can't we just hear the wisdom of our forefathers: "Birds of a feather flock together?"

UNCONSCIONABLE FREE TRADE

In light of such an upset of time-honored values and relationships, it was distressing at the least and unconscionable at the worst. It is no wonder that Marx cried foul! For Marx and Engels, industrialization meant the rape of the proletariat (the common people) for only one ultimate reason—exploitation. Marx, being an atheist, felt freer than most to speak out and lay part of the blame at the feet of the religious community. He strongly criticized the clergy of his time for not speaking out against such inhumane acts. He also blamed the crystallized political system of his time for not stepping in to change it. Economics was not his bag but he had the courage to call a spade a spade.

THE IRREVERSIBLE PROCESS

The truth is that once industrialization began, nothing could have stopped it. No one was able to foresee the ultimate results of these new technological discoveries. The humanities had developed primarily around religion, philosophy, education, and politics. Little time had been spent trying to discover man's basic relationship to the realities of economics and the creation of wealth. Strange as it might seem in our "advanced" postmodern age of today, this correlation is still one of the least understood!

The church, one of our greatest social institutions, is still very quiet about the virtues of wealth and how it is created and even less about its equitable distribution. The church of our postmodern era is focused primarily on membership growth or not losing what they have. Generally speaking, there is little interest in understanding the virtues of God's common grace to all of His creation. No attention is given by them to the labor and the intangible elements of the Personal Capital Compound of their members. And even much less in how God has blessed us all to be elements of great productivity in His kingdom.

THE NEW REVOLUTION

Over two hundred years after the beginning of the Industrial Revolution we a

re still stumbling around trying to get our economic bearings. Will the history of economics repeat itself, as it has so often on other issues? Are we not at this moment standing on the brink of yet another dynamic revolution in our relationship with technology?

Indeed we are, only this time we are not dealing with the cumbersome and bulky power of steam. Nor are we even dealing with the secrets of internal combustion, or aeronautical engineering, or electrical power, not even nuclear power. What we are faced with now is the awesome transition into the power of artificial intelligence! Artificial Intelligence, or AI, has the power to utilize and synthesize all of our present technological resources at once and to do so automatically—and then to *learn* from the experience. Currently this is in its beginning stages but it is only a matter of a few—very few—more years until it will mature into a tremendously efficient force to be dealt with.

DANIEL'S VISION

Many philosophers and Bible scholars believe that in the early centuries of human existence God, in his infinite wisdom, deliberately withheld certain areas of knowledge from mankind, especially in the areas of the mechanical and electrical power sources necessary for our present state of applied technology. Several Biblical passages support this argument, and perhaps provide us with some explanation as to why it took man so many thousands of years to advance from animal and slave power to mechanical and electrical power. One of the earliest recordings to indicate that knowledge would become accelerated in a later period was the account written by the Hebrew prophet, Daniel.

Daniel was taken captive and transported as a royal hostage from Jerusalem to Babylon in 604 B.C. Because of his talents at court, Daniel rose to a high position of responsibility under Nebuchadnezzar, King of Babylon, and continued to serve further under the reign of Cyrus, King of Persia, who conquered Babylon in 539. Daniel provides an account of the divine revelation that he attributed to have come to him through a vision relative to the succession of the nations on earth. In his prayer, prompted by the earlier writings of the Prophet Jeremiah who had warned Israel that the Temple of Solomon would be destroyed and they would be carried away into captivity, Daniel reflected his concern for the future of Israel. In response to Daniel's prayer seeking information concerning the fate of Israel he received a lengthy revelation. We record two verses for consideration of God's influence on our Personal and Joint Capital Compounds.

> Those who are wise will shine like the brightness of the heavens, and those who lead many to righteousness, like the stars forever and ever. But you, Daniel, close up and seal the words of the scroll until the time of the end. Many will go here and there to increase knowledge. – Daniel 12:3-4

FIRST BIG BREAKTHROUGH

It was over 2,300 years from the approximate date of Daniel's writing until the year 1763 when James Watt began his experiments with the best-known of the early steam engines, the Newcomen engine. The Newcomen engine was designed only for pumping water out of mines and at that time had no manufacturing applications. Watt not only perfected the design but made it three times more efficient. This increased power and efficiency paved the way for the manufacturing advances that became the great Industrial Revolution.

The most amazing irony of it all is that the principle of steam as an artificial power source was first proposed by Heron, a scientist living in Alexandria, Egypt. Heron described and illustrated a basic working model of a steam jet-propelled rotary machine in the year 120 B.C. The extreme time lag which was common between technological insight and application during this early period is dramatically illustrated in this example. It took an additional 1,800 years before Watt's creative initiative and spiritual inspiration were allowed to discover the practical applications of the basic principle. Compare that with the lightning speed with which technological discoveries are applied today! Daniel's vision takes on great credibility in the light of such a comparison.

SUPER-MICROSCOPIC CHIPS JUST
AROUND THE CORNER

Following is a dated but important source of information from the Research Institute of America. I quote from the Sept. 1983 issue of *How To Participate In The New Science Explosion*:

In the May-June 1982 issue of the *Bell Systems Technical Journal*, Bell Labs announced a revolution in chip technology. Scientists T. A. Fulton and L. N. Bunkleberger succeeded in fabricating a 'Josephson junction' chip. A Josephson junction is a minute 'sandwich' of microscopic metal layers separated by an insulating substance. When chilled to very low temperature, the junctions become superconducting, transmitting electrical pulses at accelerated speeds. Furthermore, the minuscule chips use little power and gives off a fraction of the heat of today's silicon chips. Thus Josephson junctions may be tightly packed. According to Bell Labs, a book-sized computer made with Josephson junctions could do the work of today's largest room-filling scientific computer . . .

. . . But Josephson junctions are not the only breakthrough in chip technology. Scientists are also working on chips made of gallium arsenide—a compound which gives off a minute fraction of the heat of silicon. Gallium arsenide chips would have performance characteristics similar to those of Josephson junctions, but would not require super cooling. According to industry watchers, the Japanese favor the gallium arsenide approach, while IBM is experimenting with super cooling . . .

. . . An even more incredible development is the biochip. This is the brainchild of EMV Associates of Rockville, Maryland. E. F. Hutton is arranging financing for the firm. Biochips, which could become commercial in the early 90s will be made from protein-based units that function like transistors but are much smaller and are triggered by enzymes rather than electricity. According to developer James McAlear, a network of biochips the size of today's 64K chip could hold an astounding *sixty four trillion bits of information,* and switch *100,000,000 times quicker.* "At these circuit densities," McAlear claims, "all the data stored in all the world's computers today could be stored in a single biochip computer."

In plain English, they are saying that our computer capability will increase *billions of times* with just the perfection of one biochip computer! No one can fully appreciate or imagine the impact that such a new level of artificial power will have! Nevertheless, one thing is clear: sharing in the *ownership of the application* of such technology is of great importance for the American worker to realize personal compensation for their total Personal Capital Compound before they are totally replaced by such technology.

As we look around us today we see that we are living the results of those predictions. We are using cell phones with more power than a room full of computers in the 80s. Our PCs and our laptops are also examples technological advances unheard of for those of the 80s. Our international space station and the Hubble telescope with all of its latest improvements continue to break new barriers in what we know and how fast we know it.

THE DANGER OF NOT KEEPING UP

The Josephson Junction chip, the Gallium Arsenide chip, and the Biochips are the foundation technology for what are termed fifth generation computers. Fifth generation computers are the smart computers, the ones that can *think,* and learn from experience. They are not mere "data crunchers," but are information processors. They are intelligent. Thus, fifth generation computers are the first of what promises to be a long procession in that new power source called artificial intelligence.

A practical example of this in a small way is the Super Blue computer used to play and beat the best chess champions in the world. Another more complex example are the smart bombs and missiles that, once targeted and released, can proceed to a target and—should the target disappear or otherwise give evidence that it should not be hit, such as an identifiable "friendly"—self-destruct or, in some cases, automatically return to base.

AI represents a totally new instrument for our Capital Compound—a tool so powerful and awesome the American people simply cannot imagine its effect at this point. The problems suffered by the craftsmen, tradesmen, and laborers of the pre-Industrial Revolution will seem humanitarian compared to worker displacement by the total automation capabilities of AI. We can accept the realities of this new instrument and utilize it for the good of all, or we can hide our heads and get run over.

The time is now—AI is here and won't go away. There is not a skilled job or trade that cannot be invaded by a "smart robot" or "smart machine" of some sort. Furthermore, almost all of the managerial positions now occupied by people are well within the capabilities of this technology. In the light of these demonstrated realities we must conclude that there is no "job security," there is only the possibility of "equity security" in participating in the ownership of such technology and many other instruments of our Joint Capital Compound!

In fact, the Obama administration has actually triggered an acceleration of more technology efficiencies. The real estate crash, basically caused by government intervention in the form of subprime loans to unqualified home buyers, was just another way for government to expropriate our Personal and Joint Capital Compounds in the name of Progressive Socialist policies. Obama's answer to the problem of past government intervention was predictable: more government intervention.

So now we see Obama coming to the rescue with bank bailouts, auto manufacturing takeovers, takeovers of other financial institutions, and Wall Street interventions. Then Obama announces his plans to levy special taxes on bank profits and to control executive compensations in large corporations. Obama has done all of this with impunity, as though he and his advisors are not subject to the laws of the Capital Compound Theory of Value.

The laws of our Personal Capital Compound and the Joint Capital Compound dictate that individuals and companies take measures to protect themselves from invasive taxes and bureaucratic controls. In order to do this the first rule is to economize.

When the Joint Capital Compound sets in motion its focus on economization, there are two primary places it looks to first.

1. How can I produce the same with less labor?

2. How can I produce more with technology?

Obama and his Progressive advisors told the American electorate that these new spending programs and these new controls on industry would initiate a new era of prosperity. Unemployment was to remain at eight percent or lower, and new jobs would be created in the green industries.

THE CAPITAL COMPOUND IN EFFICIENCY MODE

Before the ink was dry on the first bailout money, even before the President's inauguration in 2009, businesses all across America were cutting back on their use of personnel. Layoffs were in high gear and management, both small and great, were asking their IT contacts "What can we do to make this or do this faster and with less labor?"

Now we see what happens when the synergism of the Personal Capital Compound and the Joint Capital Compound kicks in.

Many of the medium to larger companies already had improved technologies in place but had not reduced their workforce, because they don't like to put people out of work. If the production can pay for it all and there are no serious obstacles in the way for the company or the entity to progress, then every effort is used to keep people employed. There is a strong bond between most companies and their employees, especially small to medium-sized companies. Remember, most of the employees of America *do not* work for major corporations. In fact it has not been the layoffs from the major corporations that have overwhelmed our unemployed list. It is the layoffs from the millions of small businesses that have caused this historic situation. Remember, small business accounts for seventy percent of our GNP and ninety percent of all new jobs. We only have to reverse the equation to see the impact small business has on unemployment. Obviously, if small business is responsible for ninety percent of all new jobs, then they are also responsible for ninety percent of all new layoffs.

How are we doing with that "change we can live with?" Hello, Mr. President!

THE INVOLUNTARY SERVITUDE TO WAGES

The realities of these technologies only serve to further emphasize the importance of our ability to control or to own one's own Personal Capital Compound and our contribution to the Joint Capital Compound. Anything less, to whatever degree, is an act of expropriation and to that extent slavery! A wage paid in return for only the least important element of one's Capital Compound ultimately leaves one impoverished and totally without benefit from the more important elements, creative initiative and spiritual inspiration.

In addition, even more valuable than the elements of the Capital Compound themselves, is the bonus product of synergism when these three aspects are invested in the project or company for which one is working. One will receive no compensation for any of them unless they can claim some control and ownership to their rightful portion of that entity. A mere wage will not compensate for the input of the intangible elements and their synergistic effect. Unless such compensation is realized, no matter how fractional the portion, then the more dynamic portion of one's Capital Compound has been expropriated and one is reduced to no more than a hired servant.

Understanding this and developing new ways to include all who work for a wage into the benefits of the complete Personal and Joint Capital Compounds should be one of our primary goals at anytime that we are in a position to do so. This is one of the areas of economics and management in which we are woefully behind the power curve. To say it another way, we are still in diapers on this one.

THE CAPITALIST EXPERIENCE FOR ALL

Sharing in the ownership of the applied technology of robotics and artificial intelligence is the hour Aristotle dreamed of. He wrote that there would one day be a power that would allow "chief workmen to not want servants, nor master's slaves." Now is the time. Man can now own "servants" and "slaves" without the need to restrain the personal freedom of any other human being.

In their book, *The Capitalist Manifesto,* Louis Kelso and Mortimer Adler bring out many creative thoughts and ideas for realizing such noble goals. One passage in particular is appropriate to recall:

> The conception of the machine as an inanimate slave is a familiar thought in our industrial society. But the implications of this idea are seldom, if ever, followed through to their ultimate conclusion, which is that, like the few who were slave owners in the past, it is now possible for all men to be economically free by acquiring property in the automated machine slaves of the future.

> Socialism believes that men may be free by making power-driven machinery the slave of mankind.

> But despite the fact that the economically free men of the past derived their freedom from owning capital, often including slaves, Thomas as a Socialist believes that universal freedom—economic independence and security for all—can be achieved without the private ownership of capital. On the other hand, in a recent speech, Roger Blough, Chairman of the Board of the United States Steel Corporation, cites a reference by the London economist to machines as "inanimate slaves." He recommends multiplying them in order to produce more and to distribute more widely the greater wealth produced in the form of a higher standard of living for all; but he does not implement and expand this recommendation by proposing to make all men free by diffusing as widely

as possible the individual and private ownership of our inanimate slaves.

ACCESS TO EQUITY PRODUCTIVITY IS A GOD-GIVEN RIGHT

It is important to focus attention on the fact that Kelso and Adler did not mention anything concerning the intangible elements from which all technology is derived. This is simply because, for whatever reason, neither one had ever considered such technological developments as being associated with a gift from God. To Kelso and Adler the evolution of technology and the need or advisability to share it with workers is simply a means to an end. Simple enough right?

No, wrong! There is a distinct and very important difference in recognizing the advisability of a new financial instrument—such as the ESOP for which Kelso is famous—and the understanding that the product of our Personal and Joint Capital Compound is the crucible of our inalienable right to "life, liberty, and the pursuit of happiness."

These we have as a gift from God, whether we recognize it or not. Anything less leaves open the door for others or governments to step in and say that we have no right to share in the equity creating portion of our Personal and Joint Capital Compounds. Without this right, there is little or no substance to the hope offered in "life, liberty, and the pursuit of happiness." With no right or freedom to our portion of the wealth we have singularly or jointly created with others, we are left with only wages and therefore we will have allowed others to expropriate the productivity of our Personal and Joint Capital Compound. This is not what our founding fathers had in mind.

It took animal power, servant power, and slave power to provide the driving force for the many cultures and societies of the past, to make it possible for the "serious men of property" to busy themselves in the leisure activities of philosophy and politics. For our part, we are heirs of both the positive and the negative rewards of all their efforts. We now stand at the door of an impending technological impact that will dramatically affect our lives in every way imaginable, even to the possible loss of our present freedoms. However, I do not believe this is necessary, nor am I so bereft of faith in God that I do not believe He will bless us with the answers to overcome this great new challenge we face. The evidence is clear that the Personal and Joint Capital Compounds do exist, that they are real, and that they work! The only struggle between ideologies of Progressive Socialism and the Capitalism of the Capital Compound Theory of Value is who is going to control or own it and who will be the servant or the slave?

CONTROL YOUR CAPITAL COMPOUND

I say let the instruments and the tools of the Capital Compound be the servants and the slaves, and leave our Capital Compounds free! We the people are the living organic expression of the Personal and Joint Capital Compounds. Governments, states, ideologies, religions, or corporations cannot own our Capital Compound without impoverishing us to slavery! Total ownership or control by someone else of my Capital Compound means I'm a slave, or certainly no better than a servant.

It is precisely this lack of personal, individual ownership and control in one's economic expression of life, liberty, and the pursuit of happiness, that keeps the Communist societies, the Socialist societies, and even the so-called "Capitalist" societies from experiencing the goals they claim to seek: namely, the fullest development of the whole person as a unique creative individual, and the liberation of mankind to higher pursuits other than mere materialistic production for enrichment of only a few.

America is in a very unique position in economic history to pioneer the final steps in bringing such participation in ownership to all Americans. It holds this position because of the extremely strong base of small businesses that control the largest segment of our economy. Most things grow from the *bottom up* not from the *top down*. We could see some major corporations leading the way in this pioneering endeavor. However the opportunity for success in doing so is much greater among the millions of small business owners and their families of employees. The big change will come through small to medium-sized businesses!

MAN CANNOT LIVE BY BREAD ALONE

Mankind is weary of all of this endless production and yet he wants to survive and progress. It is his most basic drive, but it does not have to be satisfied in the role of a servant or as a slave to be used only as an inanimate object would be used and considered not much above an animal (or worse, a machine) only to be discarded when no longer needed, or when a better substitute is developed. As a result of this all too common employment situation, there is deep frustration in almost all Western countries. And, although there are no official statistics available to the west, it is evident from other indicators that the Soviets and all of the Socialistic and dictatorial and semi dictatorial countries are suffering even worse. A report February 15, 1984 report in the *London Times*, dated but pointed, had this to say:

> Less than one in four Americans works at capacity. The work ethic is declining among West Germans and Japanese. And Britain has a large number of dead-end jobs.
>
> Those are among the findings of a "Work and Human Values" survey published by the Aspen Institute, an American think tank, in conjunction with European researchers and business leaders.
>
> It said unemployment in the industrialized West will continue at what used to be unacceptably high levels, and recommended new job strategies to take advantage of untapped human potential.
>
> "Unless we do this, Western democracy itself will be threatened," the 145-page report said.
>
> The survey covered 4,000 employed people in the United States, Britain, West Germany, Sweden, Japan and Israel. Here are some of the findings, which in some cases do not include all six countries because the same questions were not asked everywhere:

The work ethic, or inner commitment to work, was highest among Israelis, fifty-seven percent of whom rated it "strong." In descending order, the others were the United States at fifty-two percent, Japan at fifty percent, Sweden at forty-five percent, West Germany at twenty-six percent and Britain at seventeen percent.

Despite their stated commitment to work in general, American jobholders are using what the report calls "discretion" to hold back on their specific jobs. It said they "are choosing to give less rather than more to their jobs." Only twenty-two percent of American workers say they perform to capacity, and forty-four percent say "they do not put a great deal of effort into their jobs over and above what is required," the survey said.

About half of the jobholders said they experienced a mismatch between their values and their jobs. The "good match" percentages by country were: United States forty-nine percent, Israel fifty-five percent, West Germany forty-six percent, Britain thirty-six percent, Sweden fifty-one percent, Japan thirty-two percent.

Germany and Japan have increased the average worker's productivity impressively, while U.S. productivity growth was zero from 1973-80. But Germany achieved the growth by reducing the overall number of jobs, whereas America created one new job for every existing three.

Young Japanese workers are more committed to themselves and their families than are older workers who willingly give extra labor. "The young consider this a waste of their lives," the report said, adding that "some Japanese observers describe this as the 'Europeanization' of Japan."

The percentages of people who felt they had dead-end or bad jobs were: Israel six percent, United States nine percent, Sweden fourteen percent, West Germany twenty-one percent, and Britain twenty-two percent.

In light of our last two years of serious recession and predictions that America will not see her previous levels of employment for another decade (if then) it appears that this study, now over twenty-six years old, was a discovery of serious hidden tumors in our economic fabric which are now in full view.

SUMMARY

THE VISION OF OUR FOREFATHERS CAN LIVE ON

Despite these dreary findings, America more than any other country has the ability and the opportunity to understand and fully utilize the Capital Compound Theory of value and to set a worldwide example. The moral fiber and spiritual sensitivity and inspiration that began our pilgrimage in this great land can also make the critical difference at this time of enormous conflict and change.

Fortunately some individuals, plus a few corporations and institutions along with a number of small businesses, are making positive efforts in this direction. Yet, despite these efforts, the majority is not involved. The general level of awareness of the real problems facing us is extremely low. Financial and economic illiteracy are on a broad scale and is at the root of such apathy and lack of involvement.

This financial and economic illiteracy that we face is the root cause of our having accepted the wrong advice of the Progressive Socialists. Once one accepts such advice, their whole world view changes and their assumptions of what should be and what should not be are completely out of harmony with reality.

By the grace of God, America is coming out of her sleep. There is a new awakening to the truth and legacy of our fore fathers and to the real principles of True Capitalism, with faith, hope and love for all!

Chapter Five

Current Distribution of Wealth

Nothing is more controversial than the issue of distributing wealth. We can see this in the many sad fights and conflicts over just an inheritance. Yet this pales in comparison to the distribution of wealth in the workplace or the profits made in a transaction, where there is a serious dispute as to who gets what. This chapter reveals the history of the distribution of wealth and how this helps us to understand our own challenge for distributing the productivity realized from our Personal and Joint Capital Compounds.

GREED AND PRIDE

During the 80s, there were an estimated fifty so-called "wars of liberation" being waged at any one time, in as many different parts of the world. In almost every instance these are wars that are being fought between two basic elements, the "haves" and the "have-nots."

Most Americans find it difficult to identify with that kind of war. In America it is generally held that, if you consider yourself a "have-not," you should educate yourself, change jobs, join a union, or appeal to a government agency or other institution for help. While perhaps no one believes that our present system is the perfect solution to the problem, it is foreign for us to relate to the idea of joining a guerilla band in armed conflict against our established order! However, the

Progressive Socialists are fully capable and so ideologically driven that it is within their scope of common sense to push the country's faithful citizens to force of arms.

Equally important is the reality that in the twenty-first century we face an additional set of rebellions even more grizzly than those fought over who controls the Capital Compound. We are now faced with those who will die at the drop of a hat for ideologies that are completely foreign to us. These are the many Muslim factions who are totally committed to the destruction of all democracies, all republics, all inalienable rights to life and the pursuit of happiness, so that all can be replaced with Muslim dictatorships using Sharia Law. They have publicly stated and have written volumes to warn the world that they will kill any and all who do not agree. Moreover, they also demand complete surrender to the Muslim faith or to suffer death. Fellow Americans, these are not idle threats, these people are serious and we must also be just as serious in our commitment to not allow this to happen.

TWO IMPORTANT AND RELATED QUESTIONS

If the Capital Compound is at work in some degree of efficiency everywhere in the world, why is it that only a few countries have been able to fully develop it and exploit its productive capabilities with some degree of justice for the majority? On the other hand, why have so many others—even great populations like Russia, India and China—not been able to do so although resorting to drastic slave-like control of their people? Although, in the case of China, she is increasing in her success with the Capital Compound and the Joint Capital Compound to the degree that she is also has the fastest growing Christian community in the world. We will see much more from China in the years to come!

While the answer to both these questions is simple, the solutions are less so. The primary difficulty in both cases lies in those two most insidious human weaknesses that we have mentioned before, greed and pride! When one controls the Capital Compound, whether one is a corporation, an individual, an institution, or a government, one becomes the capitalist in the true sense. Pride of possession and power, and greed for more of the same has no bounds unless restrained by some counteractive force.

The stories of gold miners drowning in streams they were trying to cross, refusing to let go of the gold that was carrying them to the bottom, are illustrative of how powerful such forces really are. In view of the very few creative alternatives being applied in most countries for the more efficient use of the Joint Capital Compound, it would seem that the leaders and major controllers of most countries frequently resemble the drowning miners.

COMMUNISM VS CAPITALISM

We mentioned in Chapter Two that most if not all of our present world conflict and tensions were related to differing opinions concerning two major ideologies, Capitalism and Communism. Even though we have shown how both ideologies create their new wealth in the same way— through utilization of the Capital Compound Theory of Value—the two ideologies are of course quite different. Aside from many other differences between these two, the main economic difference between these two is in their method of and belief in the *distribution of wealth*.

COMMUNIST METHOD

Stated briefly, the Communist method of distribution is based on need, rather than merit. According to Lenin's interpretation of Marx, in *State and Revolution*, the Marxist principle—"from each according to his ability, to each according to his needs"—firmly replaces and transcends (in the minds of the Communists and Progressive Socialists) all considerations of justice and individual rights!

It is very important for the American worker and citizen to realize that, according to Lenin, the Communist ideal will not be fully realized until the bourgeois institution of private property is abolished *worldwide*. Can you live with that kind of change?

This sounds like "it's good to spread the wealth around." Obama believes in taking from those who have more "ability" and giving it to those who have more "need." I get the feeling that Obama and Lenin would have made great soulmates.

The Communist servant of the state (we can't really call one a "citizen" who has no voice in his society's actions) receives compensation for labor theoretically in two forms: (1) Income to cover the cost of needs, and (2) community services. In actual practice, more attention is also given today to merit, performance, and rank than was advocated by Lenin and Stalin in the early days.

Since the fall of the Soviet Evil Empire, as Reagan called it, the country has fared better but is still greatly hampered by heavy bureaucracy and the same mistaken concepts of how wealth is created and how to equitably distribute it. They do not recognize the Capital Compound, adhering solely to the Labor Theory of Value.

CAPITALIST METHOD

Capitalism, as generally practiced, theoretically compensates its citizens in four ways:

1. Income from privately-held real estate property and property in the form of Capital Compound instruments; stocks, bonds, partnerships, REITS, Sole Proprietorships, and ESOPs, plus many other similar instruments.

2. Wages based on performance and rank

3. Welfare payments for subsistence needs

4. Social and community services

Both ideologies are grossly imperfect in realizing their stated goals. However, it is obvious that the most striking difference is in the area of number 1. The other areas (2, 3, and 4) look pretty much alike, with only minor variations. Actually, in America as in all modern noncommunist western countries, we employ a mixture of the Socialist aspects of the Communist method of distribution and our own unique method of ownership of private property. *It is this mixture that is causing our biggest problems,* especially in the equitable distribution of wealth. We will examine these aspects of our own situation more closely later in this chapter, but first let's consider what the poverty-ridden peoples of Mexico, Central America, South America, Africa, India, Asia and other parts of the world see when confronted by these two ideologies.

DESPERATE PEOPLE

First, one has to accept the fact that these poverty-ridden people aren't really all that interested in studying ideological principle. Generally speaking, the one and only concern they have is for their own desperate personal situation. We are so blessed in America that the great majority of us have no idea of the real meaning of the word desperate. Desperate in its truest human sense, describes a life-threatening need. For most Americans, desperate is a situation where one must watch the Super Bowl on anything but a wall-mounted HDTV!

Josephus, the Hebrew scholar and historian who was captured by Caesar's army in the first century A.D., wrote concerning the tragedies associated with the fall of Jerusalem. He stated in his writings that, among the many tragedies of the day, one stood out above all others. After nearly two years of siege that had held the city of Jerusalem in an iron grip, two Jewish women were reported to have reached such a state of despair that they resolved to kill and eat their own babies. One proposed to kill and share her baby between the two if the other would do likewise. They reportedly agreed.

Desperation is not interested in ideologies, desperation wants results. When the level of desperation reaches a certain critical point, those affected no longer react in ways that we would call rational.

EVIL ACTIONS CAN SPEAK LOUDER THAN WORDS OF INTEGRITY

Consequently, when poverty-stricken people in any society are given the opportunity to choose ideologies, their first question is most likely to be "What does your cause do for me now?"

Ask yourself this question. What would you do? If you found yourself in such a desperate situation of poverty, which system of distribution would you choose? Would you choose the one that says, we are for your personal freedom and we will give you the right to be independent and an opportunity to vote for your leaders, with the privilege to keep living basically as you have in the past?

Or, given the opportunity would you choose the one that says, we will help you get what you need now if you will join us. We are going to take control of all the property of this country and see to it that the wealth is properly divided among all of the people?

I'm pretty sure which one I would be most tempted to choose, how about you?

Now, when all was said and done, I might be very sorry that I accepted the latter offer. But, in the absence of any "better" offer being available to me at the time that would address my immediate suffering, my choice would seem very rational. Relative to my state of desperation, what realistic alternative would I have? The motivation for the choice thus made is really not between ideologies, but rather the immediate need for a distribution of some wealth in order to live. The later would take too long and address only my long-term needs. But, at that time, I could only see and feel my short term needs. It is greatly to be despised , but a reality for many unfortunate souls in our current world situation.

So, we see that while all cultures and societies create their new wealth through utilization of some form of the Capital Compound Theory of Value, not all feature the same methods of distribution.

USING THE EXTREME TO UNDERSTAND THE SIMPLE

Most who read the above scenario would likely say to themselves: "Wow that is a bit extreme. I can't imagine myself and my family ever experiencing anything like that." Ok, I feel the same way. However, when we refocus the above scenario into our current situation in America, the extreme might prove to be a teacher for us to learn something simple but not easy to perceive.

Allow me to restate the second part of the scenario: Given the opportunity would you choose the one that says, we will help you get what you need now if you will join us. We are going to take control of all the property of this country and see to it that the wealth is properly divided among all of the people?

Obama says, "We will help you get what you need now, just vote for me. I will take control of all of those who produce evil profits, such as oil companies, banks the pharmaceuticals. I believe in sharing the wealth with those who deserve it most."

Now, which one did our nation chose in the 2008 Presidential elections? We all know the answer to the question: we chose the Senator from Illinois, Mr. Barack Obama. It would be easier to swallow if we could convince ourselves that we did not really know what he stood for or what he would actually do.

The truth is, we can't hide behind that excuse. We all heard him tell the plumber that he thought it was a good idea to spread the wealth around. We all heard him say that he was going to increase taxes. We all heard him say that many companies had "excessive" profits. (Read that to say, evil profits.)

So, we can't hide, but we can admit that we took the selfish way out. We accepted the idea of stealing from those of greater ability and giving to those of lesser. It's stealing whether it is taken from an individual or from a large company. The money does not belong to the recipient of welfare. It belongs to those from whom it was stolen. Oh, how slippery is the slope that takes us down the road to serfdom! America, your reality is upon you!

THE AMERICAN EXAMPLE

As a people of faith with hearts of mercy and personal concern for others, there is no compassion anywhere in the world that can equal the American people. From the earliest colonial times, Americans have always been a people of faith and mercy for the less fortunate. Private assistance by the American people is measured annually in multiple billions of dollars. All of the countries combined in the world do not rise to the annual donations of money and material assistance made by the American people—for the benefit of others—nationally and internationally.

The concept that government is the best source of assistance to the unfortunate and those in need is a page out of the Progressive Socialist playbook. In the pure application of the Progressive Socialist doctrine as originally applied in Russia, China, Germany under the Third Reich, Spain under Franco, Italy under Mussolini, all the Muslim Middle Eastern states and many others, *private* assistance is discouraged and in many instances it is forbidden.

Giving assistance is considered by the Progressive Socialist movement to be an act of divine nature. It illustrates power and the ability to influence others through gifts. Pure Progressive Socialism wants and even demands that all assistance come from or through them. All of our tax increases this year and next are designed to spend your money on Obama's Progressive Socialist dreams. Are we enjoying the change yet?

CURRENT EVENT NOTICE

On March 19, 2010, the most popular and successful political commentator in the world today, Rush Limbaugh, had this emotional reaction seen here in his quote to Obama's latest Progressive Socialist move in health care:

> Like you, I am shocked! I am infuriated by the Obama-Care scam of Nancy Pelosi. This is the overthrow of the U. S. constitutional system! I have had enough, actually too much! Words rush to my mind trying to describe Barack Hussein Obama's first year as President, words such as, recklessness, folly and failure.

Whatever words you use, "We the People" find ourselves on the precipice of either victory or defeat! America is literally moments away from a National Socialist overthrow of our beloved nation or the resounding defeat of the Obama-Reich.

This is nothing more than a 1927 style Putsch, forcing our nation into the lustful hands of the anointed few.

The Obama coup d'état is being finalized as you read this! The liberal cabal is still trolling for votes and if enough votes are not secured then Pelosi and the members of the Obama-Reich will try to force the National Health-Care scheme into law, but you and I are in the way!

THE CUTTING EDGE

Speaking in economic terms, distribution is the cutting edge of reality. It is where the "rubber meets the road." The distribution of wealth touches the emotions and can create an atmosphere of tension or controversy more quickly than nearly any other aspect of our lives!

Yet, equitable distribution is not impossible. Intelligent men and women, by taking proper steps, are making equitable distribution more certain in their lives. Our challenge is not only that we should achieve creative methods of dynamic, equitable, and stable distribution of wealth for ourselves in America, but that we should offer the same opportunity to all the citizens in any other country, where possible.

NO PERFECT ROLE MODEL

We must keep in mind the fact that no culture or society exists today or has ever been found that has developed a method to perfectly employ the Capital Compound Theory of Value to its maximum efficiency of operation. That is to say, one where all are able to have everything they need and some to spare, where control of the Personal and Joint Capital Compounds are enjoyed equitably by all, and where each receive just compensation for participation in the creation of new wealth and receives equal benefit from the wealth of societal communities at large. The challenge is still with us!

This is precisely what Marx and Engels thought their Communist philosophy would do for mankind. Marx wrote that:

> "The division of labour offers us the first example of how, as long as man remains in natural society, that is as long as a cleavage exists between the particular and the common interest, as long therefore as activity is not voluntarily, but naturally, divided, man's own deed becomes an alien power opposed to him, which enslaves him instead of being controlled by him. For as soon as labour is distributed, each man has a particular, exclusive sphere of activity, which is forced upon him and from which he cannot escape. He is a hunter, a fisherman, a shepherd, or a critical critic, and must remain so if he does not want to lose his means of livelihood; while in Communist society, where nobody has one exclusive sphere of activity but each can become accomplished in any branch he wishes, society regulates the general production and thus makes it possible for me to do one thing to-day and another tomorrow, to hunt in the morning, fish in the afternoon, rear cattle in the evening, criticize after dinner, just as I have a mind, without ever becoming hunter, fisherman, shepherd or critic. (Marx, Communist Manifesto)

What an idyllic vision of the future for the Communist man and woman! In addition to being idyllic, from a workers point of view, it was even democratic. Why else would one become a critic from time to time? Certainly *not* to be sent off to Siberia, which is generally the fate of those in Russia who dare to criticize. Marx never seems to quite get around to describing the infrastructure necessary for "society to regulate," as he put it. The concept was that the state would somehow pull off whatever miracles were necessary to make Marx's philosophy a reality. The whole thing seems to be taken as a matter of course. Overpowering ideology needs no logic!

COMMON SENSE OR NOT

Obviously, what has taken place in the course of history since Marx wrote those words, so full of hope, is something other than what he had in mind. The great concentration of control over the Capital Compound in the hands of the state has just not worked out as planned. Instead of the idyllic paradise of which Marx dreamed, something much different has evolved.

When centrally regulated (and thus suffocated with detailed regulation of every sort) the Capital Compound loses the creative initiative to function productively. Oh, things go along for awhile, as long as the ardor of the group and enthusiasm for the goals are kept alive and warm. But after many years of continual struggle for the benefit of the state, the vision dims and altruistic desires subside.

Oh, that the above were absolutely true! Alas, it is not. Unfortunately, in the world of Obama and his Progressive Socialist advisors, the dastardly lie is alive and well!

Tell me, how is it possible for the very country which called out the evil Russian Empire—challenging it to "put up or shut up" as Reagan implied when he yelled at Gorbachev, "Come down here and tear this wall down"—to be so gullible? How is it possible that the very same country, in less than twenty-five years, can be on the brink of accepting the Communist ideology as our new model for the ideal America? Quite frankly, it is *we the American people*.

We have made it possible simply by looking the other way. As a republic, given to us by the grace of God, we control who rules over us and the laws that they pass and the responsibilities that we allow them to assume. We did not get here as people controlled by a dictator. If that were so, it would give us an excuse to say we were innocent victims.

On the contrary, we have self-consciously stood by and allowed it to happen. Because of our own problems with personal greed and our cumulative lack of knowledge of the truth, we have voted into power individuals who have the ultimate destruction of our beloved America as their sole desire.

Looking back from 2010, the year of this book, it has been over ninety-two years since October of 1917, when Lenin rose to power as the Supreme Leader of Russia and the new Communist state was launched. In that short time the whole world has become committed to one form or another of Marx's philosophy of the Labor Theory of Value. All struggled gallantly to make it work. But try as they will, those who adhere closely to Marx's theories find that, inevitably, they must come to the bargaining table of trade with those countries of the world who still allow significant freedom for private ownership of property and some level of free trade.

This does not keep those who have committed themselves to the Communist/Progressive Socialist Ideology from believing that they will ultimately succeed. Obama and his Progressive Socialist advisors have been in training for over twenty years—just waiting—for this opportunity to finally bring all of their guns to bear on our beloved Republic and kill it once and for all. Nor does the failure of such an arrogant ideology appear to discourage those emerging countries, full of poverty-ridden people who are crying out for a solution to their problems, from running to embrace the Communist/Progressive Socialist ideals.

A DEEPER LOOK AT DISTRIBUTION

How can we effect the changes necessary to give us a more dynamic and equitable distribution of wealth than we now have and at the same time, offer the same hope to those millions of the world that are plagued with the disease of poverty?

First, let's examine more closely the way we in America are presently distributing our wealth. Then we will more naturally arrive at the solution.

In our model of the Personal Capital Compound we have the now-familiar three elements: Labor, Creative Initiative, and Spiritual

Inspiration. Although every worker in America makes some use of all three elements in their respective jobs, the great majority receive a wage based on an hourly rate. An hourly rate is a direct recognition of only the "labor" element of the Capital Compound. A smaller percentage receive income from perhaps some rental property, from cash investments in bonds, financial instruments such as the popular T-Bill, or from shares of stock that pay a dividend.

Such non-wage income, sometimes referred to as "passive" income, is a benefit of the creative initiative and spiritual inspiration elements of the Capital Compound. In the last fifteen years, the creation of various methods of investing to save on taxes, the development of limited partnerships with national syndication, and the emergence of more aggressive retirement programs for all workers have increased the distribution of Capital Compound instruments into a much broader sector of the American workforce. Subsequently, many more Americans are now receiving some portion of their incomes from the two intangible elements of the Capital Compound than was the case four decades ago. But it is far from being generalized as a practice or even as a basic understanding.

Despite these advances and the wider distribution of such instruments, we still have enormous concentrations of Capital Compound instruments in the hands of major corporations and government. To make matters worse is the fact that the majority of those who own some capital instruments could not conceivably live by the income derived from their ownership. They still need to rely on some form of work that pays an hourly wage. Does the second and third job come to mind?

Generally speaking, our present system of laws governing the distribution of profits created by the efficient use of the Personal Capital Compound does not allow for the owners of those instruments to receive the full income that they generate. There are exceptions to this generality, but I daresay that less than two percent of Americans know what they are! We will review them later in the book.

OPPORTUNITIES FOR ENTREPRENEURS

In addition to the wage rate method of distribution, America is a great haven for the entrepreneur—those gutsy few who are willing to risk all for the opportunity to own and manage their own business. From the earliest years of the colonist business owner to the present, more than ninety percent of all *new profit producing jobs* in America have been provided by these hearty people. Thank God for such freedom and for such people willing to take the risk and the advantage that such freedom provides.

After all these years, big corporations account for only thirty percent of our Gross National Production and only five to ten percent of new profit producing jobs annually. Yet it is the big corporations that represent the largest single employers along with the federal government and also the largest single concentrations of Capital Compound instruments, second only to the federal government!

Despite this, we see that it is not the big corporation that is providing the new jobs necessary for growth and, with the present trend toward larger and less efficient government, it is impossible for our government to create any positive impact on our economy through its overwhelming efforts to grow the number of bureaucrats. None of this effort produces economic profits for the benefit of the country. Government is an expense to the economy of any nation, and ours is no exception.

SOCIALISTIC ASPECT OF OUR DISTRIBUTION SYSTEM

I have stated that we have a mixed form of distribution in America, one that combines Socialism and Capitalism. It may not be easy for everyone to understand this at first glance, so allow me to elaborate.

The Socialistic aspect of our distribution is exemplified in the determination of our society to encourage full employment as the primary means of distribution. Remember, Progressive Socialism depends on full employment for ultimate success. The *Full Employment*

Act of 1946 states, in concept, that it is the accepted policy and goal of the federal government to encourage an economic environment that provides for "full employment." On the surface this seems both noble and logical and of normal national concern. After all, didn't even the Apostle Paul advise the early Christians in his letter to the Thessalonians: "For even when we were with you, we gave you this rule: 'If a man will not work, he shall not eat.'" So it appears to be a simple straightforward matter—one must work if one wants to receive any wealth, goods, or services, right?

But a problem has already arisen which will defy this concept. From the outset, "full employment" implies that our society, along with other societies which purchase our exports, will be able to consume all the goods and services that we can produce with our Personal and Joint Capital Compounds. If the rate of increase in the production capability of our Joint Capital Compound were limited to the rate of world population growth, this might appear to be a workable assumption. But reason must surely prevail when we consider the impending *explosion* of the production capability of the intangible elements of our country's Joint Capital Compound.

Physical strength, as related to "toil" in the production process, has its limits. Although our creative initiative and our spiritual inspiration are not infinite, they are much more productive by comparison. Despite this, we promote the Socialist concept of "full employment" and insist that everyone participate directly or indirectly in the production process. In reality what we must do, and that which is our great challenge, is to implement a plan to distribute the wealth generated by the nearly limitless production capabilities of technology. We can do this through individual ownership interest in the "instruments" of that production. We will detail how this could be accomplished in later chapters.

LIMITED VISION

The one basic conceptual error that has inspired and motivated both the Communist and the Capitalist ideologies to pursue a similar course relative to distribution is the concept that all wealth is generated by labor. The sole exception was our Colonial period, when over ninety-five percent of colonists were entrepreneurs through a family-owned business or farm. Therefore it should not surprise us to find that, despite our avowed beliefs in the right of everyone to hold private property, the right to vote, the right to participate in public and private institutions, and so on, we still hold to an inaccurate and incomplete view of the source of wealth.

This is how a departure from faith in the action of all three elements of our Capital Compound was begun. The resultant narrowly focused pursuit of the one element—labor—was due largely to a popular inability of everyone involved to acknowledge the dynamics of the two intangible elements. When the Capital Compound was totally in the hands of men and women who were taught from birth that they received their blessings and their wealth from God, it promoted a fierce work ethic and a heightened desire to be productive for the honor and glory of God. Compare this with our current situation of obvious financial and spiritual illiteracy, proven by the fact that we voted for a man who promised to bless us by stealing from others.

A DE FACTO CLASS SYSTEM

As so very often happens when we are confronted with change, instead of embracing the change and discovering its positive message for us, we view the elements of change as our enemy. In the same way, whenever the creative initiative and spiritual inspiration elements of the Capital Compound begin to dominate over the physical element, labor, it is easy to view those elements as our "enemy." But, unlike the change that Obama has brought, this is true change that we *can* live and prosper with!

As a result, it has been inconceivable to all earlier generations and societies that the worker should ever be compensated for anything other than labor. Ever since the earliest workers gave up their personal ownership of the land—their Personal Capital Compound—and began to serve landowners as artisans, craftsmen, and common laborers, an incorrect assumption has prevailed. Mankind has tacitly accepted that there are at least two class distinctions: "laborers" and "owners of property." Property, in this case, is more correctly defined as an instrument of the Capital Compound.

Furthermore it has become common to assume that those who work for (or exchanged their labor for) a wage or a subsistence, have no claim to compensation from the two intangible elements of their contribution. Even though the fruit of the intangible elements, of the Personal Capital Compound, can be seen in the end product of the Joint Capital Compound effort, no sense of justice has developed, relative to its disposition as an asset. Since colonial time, we simply have not been accustomed to thinking in these terms. However, in light of the Capital Compound Theory of Value, can we afford to continue to tolerate such ignorance?

If we were to propose these principles of the Personal and Joint Capital Compounds to a community of early colonists, say in Massachusetts, what could we imagine that they would say? Would they say, "You can't be serious, you must be nuts," or would they more likely say, "What is new about that?"

I personally think they would reply with the latter. It makes no sense for a people who were engaged day and night, working and understanding the strength and productivity of their Personal and Joint Capital Compounds, to think that it was anything less than "common sense." What do you think?

ORGANIZED LABOR AND THE CAPITAL COMPOUND

Unions and unionism were born out of the basic Socialistic movements. People were able to see that labor was not getting a fair shake, and some form of organization was needed to force the optimal controllers of the instruments of Joint Capital Compounds to share the wealth created with the laborers. For this reason, men banded together into groups called Collectives at first, united, or working in unison. Later the terms "union" and "organized labor" were adopted. However, we must not think that this discovery of the need to do something to control the distribution of the productivity of the Personal and Joint Capital Compounds is unique to our post-Industrial

Revolution era. Trade societies, as some were called, existed before the Roman Empire.

The unions formed during the Industrial Revolution were met with fierce resistance from both factory owners and governments. The assumption that the worker could not claim any equity or share in the instruments of production, other than whatever wages he or she was willing to accept unfortunately prevailed as an accepted universal law.

At the turn of the twentieth century, most workers (if not all) were being exploited to some degree in all three elements of their Capital Compound. While they contributed about ninety percent of the input that was needed to turn out a finished product, the pay was generally less than ten percent for their efforts. There was also *no consideration* given for any input from the creative process, or the value which could be ascribed to Creative Initiative or Spiritual Inspiration or, for that matter, the value of the synergism generated by the Joint Capital Compound of the whole workforce.

Capital Compound instruments in the form of cash, machinery, plant facilities and management contributed—in the minds of the owners and managers—the other ninety percent in the production process. The true owners of these capital instruments—whether the stockholders, founders, or managers—reaped extraordinary profits without any sense of how blind they were to the value of the Joint Capital Compound of all of those employees who made it work and to be profitable! Needless to say, the situation was greatly imbalanced and unfair.

At this juncture in the evolution of the Industrial Revolution a choice had to be made in order to keep peace. Either wages or benefits had to go up or workers had to share in the ownership of the instruments of the Joint Capital Compound. Since there was little or no insight into the dynamics of the Joint Capital Compound—and as labor, only one element of the Capital Compound, had been so forcefully presented as the only source of all wealth—it was inevitable that a limited concept of wages and benefits would be chosen as the "correct" path to follow.

Having thus chosen, it also followed that "Full Employment" would be a top priority of the society of workers, which formed the basis for our current Progressive Socialist movement.

Departing from the basic reality of the Capital Compound involving all three elements to a pursuit of only one of those elements, labor, has resulted in a myopic national focus on wages. As a result, what was earlier resisted by both government and employers came to be partially accepted as necessary for a semi-just distribution of wealth.

This acceptance was not immediate but has become very evident in both the long record of private industry's acquiescence to better wages and benefits for workers, and the government's growing support of organized labor that lasted until just recently. Currently it has now been renewed under the Obama administration, not as an effort to correct old myopic views or to create a more equitable distribution, but to justify the increase in government power and control.

The acceptance of this laboristic concept was not without its opponents. Many factory and business owners of the period spoke out for alternative methods of distribution, such as shares and profits, but they were so few that they have been drowned out by the proponents of laborism. We will look at some of these capital instruments in later chapters.

INDUSTRY TAKES A POSITIVE STEP

When Henry Ford initiated the first five-dollar workday, he at first was criticized as being pro-labor by his manufacturing peers. Later, it was acclaimed as a stroke of genius. Ford realized that in order to have a broad market for his mass produced cars, the laboring class was going to need better incomes with which to purchase them. The idea did not originate with Ford, he was just the first to put it into practice.

LABOR'S VISION OF CONSUMPTION

Union leaders clearly recognized this need one hundred years later. However, earlier, in 1827, the *Preamble of the Mechanics' Union of Trade Associations* in Philadelphia had this to say:

> If the mass of the people were enabled by their labor to produce for themselves and families a full and abundant supply of the comforts and conveniences of life, the consumption. . . would amount to at least twice the quantity it does at present, and of course the demand, by which alone employers are enabled to subsist or accumulate would likewise be increased in an equal proportion . . . All are dependent on the demand which there is for the use of their skill, service, or capital, and the demand must ever be regulated by the ability or inability of the great mass of people to purchase and consume.

Phillip Murray (1886-1952) succeeded John L. Lewis as president of the Congress of Industrial Organizations (CIO) in 1940, and held that post until his death. In his Annual Report for 1952, he said:

> Our mass production economy can expand on a healthy basis in the long run, only if it is based on rising levels of consumption of the output produced by expanding productive facilities.

These statements are the nucleus of our present day adherence to the concept of "Consumerism." On the surface, consumerism seems to be the crucible of the American dream. The truth is something quite different.

The wealth and prosperity of the early colonists, which puts our current per-capita income to shame, was not connected to consumerism. The colonists were rich and prosperous because of their ability to produce everything they needed for a comfortable living. This reality was the foundation of America for over two hundred years before Marx appeared on the scene.

THE VANGUARDS OF LABOR

Organized labor unions—with their successes in uniting the diverse aims of the wage earner into comprehensible goals of better wages, shorter workdays, and better benefits—were the undisputed vanguards that launched a growing acceptance of basic labor rights. A brief recap of our nation's history relative to the achievements of organized labor and their close correlation with political support will serve well to illustrate the relationship.

In 1792, the Philadelphia shoemakers organized the first local union. In 1799 they won the first union contract after a ten week strike.

In 1825, the tailoresses in New York City organized the first women's union.

In 1834, union groups founded the National Trades Union, the first national labor federation, in New York City.

In 1842, a Massachusetts court ruled labor unions legal.

In 1866, union groups organized the National Labor Union, the first important national union association.

In 1869, Garment workers founded the Knights of Labor in Philadelphia and the carpenters in San Francisco adopted the first union label.

In 1884, Congress set up the Bureau of Labor as part of the Department of the Interior.

In 1886, trade union leaders founded the American Federation of Labor (AFL) in Columbus, Ohio.

In 1892, strikers and guards fought at the steel mills at Homestead, Pennsylvania. Ten persons died in the struggle.

In 1894 President Grover Cleveland called out troops to keep order during the Pullman strike in Chicago. More than $80 million in property damage was reported, equivalent to approximately $8 billion in 2010 dollars. Also, that same year, Congress declared Labor Day a national holiday.

In 1913, Congress created the Department of Labor, separating it from the Department of Commerce.

In 1919, the first great nationwide wave of strikes idled four   million workers, and the Boston police force struck, making it the first strike by government employees.

In 1932, the Norris-La Guardia Act limited the use of federal court injunctions in strikes.

In 1935, Congress passed the National Labor Relations Act (Wagner Act) to protect union rights, and created the National Labor Relations Board. Also, that same year, the Committee for Industrial Organization (CIO) was organized in Washington, DC. It is now known as the AFL-CIO Union.

In 1935, the government enacted the Social Security Act.

In 1946, Congress passed the Full Employment Act.

In 1947, Congress passed the Labor-Management Relations Act (Taft-Hartley Act) over President Harry S. Truman's veto.

In 1955, the AFL and CIO merged after many years of consideration by their top leaders.

In 1959, Congress passed the Labor-Management Reporting and Disclosure Act.

In 1963, Congress passed the Federal Equal Pay Act, requiring employers to pay men and women equal wages if they perform the same duties. However women were not satisfied with the results.

The BSEIU changed its name to Service Employees International Union (SEIU) in 1968.

In 1971, President Nixon signed the Emergency Employment Act of 1971, authorizing $1.15 billion over two years to finance some 150,000 new "public service" jobs. This is how government produces jobs, but as we have already said, not one of these jobs is involved in an entrepreneurial process dedicated to making a profit.

SOCIALISTS AND SOCIALISM

In America, leaders with Socialistic concerns for the people have manifested themselves in all political parties. The architects of our Declaration of Independence, and subsequently our Constitution, were men of great social concern, and accordingly have been called "Socialistic." However, when we refer to Socialist or Socialism in relation to the distribution of wealth, we are referring to a particular conviction regarding wealth, how it is generated, and what part government should play in its distribution. Now in our days of 2010, it is more correctly stated as Progressive Socialism.

Labor leaders since before the time of Karl Marx (as well as in many different schools of politics) have for centuries have proposed various methods of overcoming man's plight concerning distribution. Socialism and Socialists, as both a movement and as individuals, generally believe in and promote distribution of wealth from a strong central control point. Yet, so-called "pure" Progressive Socialists in America do not believe that the state should own and control all the instruments of the Joint Capital Compound yet, but certainly those deemed the most lucrative. Contrary to our God-given republic form of government, they believe strongly in a democracy or majority rule as a means to gain power but they are *not* strong supporters of our republic form of government, which protects the minority against the majority. It is a matter of record that both of our nation's major political parties have political leaders of various strengths and influence, who hold strong Socialist convictions. As a rule however, they tend not to frequently use or mention the words Socialist or Socialism so as not to arouse any ill feelings or prejudice—they now refer to themselves as

Progressives. Even many politicians, whom we have always considered to be conservative, such as John McCain, are proud self-proclaimed progressives.

Despite the many democratic ideals of the Communist/Progressive Socialist ideology of Marx and Lenin, the major Communist parties maintained, at least before the fall of the Berlin wall, that socialism had already been achieved in Russia. By their definition, Socialism is but a step on the road to Communism, or total control of our Joint Capital Compound! They believe that once Capitalism has been abolished worldwide, the Communist dictatorship will wither away, and then the state itself, leaving only the pure proletariat (the people) in charge.

How is that for altruism? However, despite it's overarching optimism, it sounds just like something Obama would say.

The Socialist party in the United States was organized in the 1890s, and their membership had increased to over four hundred thousand by 1904. By 1924, when the Socialist party supported the Progressive candidate for president, Robert M. Lafollette, their voting strength had reached about one million. Since that time their voting numbers have varied greatly according to the prevailing political climate. Socialism as a political movement is closely associated in the minds of most Americans with Communism, so the party has never achieved any real power: thus their adoption of the term "Progressive."

Those politicians with strong Socialist convictions generally align themselves with the major party that could put them into office. Therefore, as a nation we have come to accept a mixture of both Capitalism and Socialism as the norm. Consequently, gradual changes in our distribution methods have followed the same developmental path. We have become progressively more laboristic in our form of distribution of wealth, and more restrictive in our Capitalistic form of distribution. Now we can more clearly understand why the politicians of the Socialist persuasion prefer to be called Progressives. The word Progressive is under the radar of the majority of Americans—that is until now!

Needless to say, relationships between labor in general, both organized and unorganized, and the persons in control of the instruments of our Joint Capital Compound have been less than friendly. The latter has come to be known as "management" or "capital." The "opposition" is divided into two camps: "organized labor" and "unorganized labor, or open shop."

Regardless of the history of how we arrived at this point, we are not likely to get where we want to go by maintaining adversary relationships! From a laborer's point of view, both the organized and the unorganized want the same financial results: control over their Personal Capital Compound!

PERSONAL RESPONSIBILITY

In order to arrive at solutions and seek creative alternatives to our present problems both nationally and internationally, we, the working people of America, must first be willing to increase our understanding of the issues. It is good to have leaders and to trust in leaders, but if the leaders are not supported by *informed* constituents, then the followers function merely as sheep, going blindly wherever they are led. Many of our present day leaders are products of a learning experience which consists of more theory than practical firsthand experience. They draw conclusions and make decisions based on imperfect studies, the advice of other world leaders with the same limitations that they have, and their peers.

Unfortunately, it appears that this limited type of learning tends to make our leaders promoters of theories, ideologies, and philosophies that they have learned in the classroom, rather than from role models in the laboratory of life. Consequently, we as a nation have been led into experimenting more with Communist and Socialist theories and philosophies rather than with the other possibilities of the Joint Capital Compound. We should be going after the promotion of worker ownership of capital instruments; worker financial independence through alternative compensation methods; the involvement of the working people in "leisure activities" such as politics, community service, and the encouragement to seek inspiration; and personal development, to name but a few.

A DEEPER LOOK AT WORLD COMMUNISM

We cannot review the developments of the last century without realizing that the control of property and the distribution of wealth are at the center of worldwide unrest and violence. So what can we learn from those who experienced the results of the Communist experiment, from those leaders who embraced the Communist ideology of total abolition of private property?

Before Yugoslavia was broken up after the failure of Communism—and although it was a country committed to the Communist ideology and philosophy—it refused domination by Russia and was looked upon by many Western leaders as an island of "Progressive Communism." Surely if a positive word for Communism is possible, it should be forthcoming from a country like Yugoslavia.

The Winter Olympics of 1984, held in Sarajevo, Yugoslavia gave much of the world the opportunity to view the outward positive appearance of this Communist state in the picturesque setting of a winter wonderland and competitive sportsmanship. In the beautiful display and pageantry of the Winter Olympics, Yugoslavia looked serene and at peace. However, facts of history have proven that it was merely a show of dictatorial showmanship put on by the government. But, what of the real heart of the people, who could possibly speak for them?

They rebelled and spoke for themselves. In the end they shot the dictator, Slobodan Milošević and his wife in the basement of their palatial residence. Milosevic had inherited the power after Tito. It was commonly reported that he died of a heart attack, however after the dust settled a documentary investigation was done and it was revealed that he was captured and shot.

A TOP COMMUNIST SPEAKS OUT

The Communist ideology professes to be so pure and so perfect that they place a very high price on the privilege of speaking out with opposing views. Perhaps a gold medal for personal courage and sacrifice should be given to a former vice-president of Yugoslavia and a top functionary in the Communist Party, Mr. Milovan Djilas! Djilas was sentenced to seven years imprisonment in 1958 for publishing his change of convictions relative to private property and the classless society.

Djilas wrote a book titled *The New Class* in which he proclaimed that the power of the bureaucrats is far greater under state Communism, because they control all capital property in the name of the state. Under such conditions, the society may be *nominally* classless in an economic sense, according to the assumption that all men are proletarian and none owns capital property. Djilas wrote:

> As defined by Roman law, property constitutes the use, enjoyment, and disposition of material goods. The Communist political bureaucracy uses, enjoys, and disposes of nationalized property. . . The new class obtains its power, privileges, ideology, and its customs from one specific form of ownership—collective ownership—which the class administers and distributes in the name of the nation and society.

Wow, I wonder what advice he could give to our illustrious President Barack Obama. Wouldn't you just love to be invited to that meeting?

A LEADING SOCIALIST SPEAKS OUT

Max Forrester Eastman (1883-1966) was an American writer who's *Enjoyment of Poetry* (1913) introduced many readers to poetry for the first time. Eastman was a strong Marxian Socialist, and was instrumental in founding the magazines *The Masses* (1911) and *The Liberator* (1918). However, as he continued to study and witness the actual application of Marxist philosophy in Russia and the Eastern Bloc countries, he experienced a major change in his convictions, similar to that of Milovan Djilas.

In 1955 Eastman wrote a book titled *Reflections on the Failure of Socialism,* in which he outlined his new convictions in opposition to

Socialism. However, prior to the publication of that book he wrote an article for the *Reader's Digest* in 1941, in which he stated:

> It seems obvious to me now—though I have been slow, I must say, in coming to the conclusion—that the institution of private property is one of the main things that have given man that limited amount of free and equality that Marx hoped to render infinite by abolishing this institution. Strangely enough Marx was the first to see this. He is the one who informed us, looking backward, that the evolution of private Capitalism with its free market had been a precondition for the evolution of all our democratic freedoms. It never occurred to him, looking forward, that if this was so, these other freedoms might disappear with the abolition of the free market.

Both of these former students and proponents of the Marxist and Socialist philosophies came to the same realization relative to wealth, power, and freedom. They made that most difficult of all transitions from proponents of central control, power, and distribution to the basics of private and individual ownership of capital instruments. Most everyone understands the importance of having quality foundations under a building or a bridge. The building or bridge may be of extraordinary design and beauty, but if the basic structure and strength of its foundations are weak and improperly constructed, then the whole structure is in great danger of collapse.

So it is with the building of financial structures. If the foundations are grossly imperfect, then the whole structure is going to become top-heavy at some future point and begin to weave and sway and be in imminent danger of total collapse. The most important building block in the foundation of a society's financial structure is private individual ownership of capital instruments. Capital instruments are the only enduring result (eternal quality) of our Personal Capital Compound contribution to the production process. The society whose financial structure does not aggressively pursue individual control and ownership of a just proportional amount of one's Capital Compound contribution to the production process is building on a financial foundation that will ultimately collapse, whether that be Soviet Russia or the United States of America!

FULL EMPLOYMENT

As we mentioned earlier in this chapter, the concept of Full Employment as outlined in the *Full Employment Act of 1946* sounds both noble and logical. However, if labor had not accepted as fact the proposition that the only way to be compensated for their input to the production process was through wages and benefits, that grand proposal would not have looked so good. By emphasizing labor and wages as our place or class, we ignored the real pool of wealth, the wealth we left behind in the creative initiative and spiritual inspiration we contributed to the production process. In addition to this, we also left behind the effects of synergism that was the gift of "added value" from all our joint efforts.

GLOBAL FORTUNES

Who are the beneficiaries of all this accumulated wealth that has not been distributed? We have only to look around. The nation's wealthiest businesses and institutions, especially government, offer ample testimony to the identity of the recipients. We now have major, and even so-called medium-sized corporations, which have budgets that exceed the annual budget of many countries of the world. The temptation is to ask the question: How is it possible that corporations, companies, and institutions can continue to amass such global fortunes in capital instruments and yet the individual workers end up with so little of lasting value?

Is the mystery beginning to clear a little? If you were in a position to receive, day in and day out, an input of financial strength and vitality of a volume that consistently exceeded more than four times what you were required to pay back out, could you not amass a great fortune? This statement might seem exaggerated to some: however, it's actually quite conservative. In areas of high technology, and even those of just simple inventive creativity, the gains realized from the "intangible elements" of our Capital Compound are often hundreds, even thousands of times more productive than the compensation paid. Should this ever be proven, how would one get compensated for their intangible contribution? The answer to that is coming.

FIXED COSTS

Assuming for a moment that what has just been said is true, why can't this wealth be shared in the form of wages, simply by letting the worker have his share in cash upfront? That's a logical question and the answer is not very difficult, considering what we have already discovered.

Wages and paid benefits, unlike profits, are a fixed cost. That is, they must be paid before the product or service can be marketed. If, for example, we calculate incorrectly and too much must be paid out in wages and benefits, we must raise the price of our product in order to gain the profit we need for the company to survive and to pay its owners. This could be disastrous if someone else is also making the same product or offering the same service with equal quality at a lower price. We are then restrained from raising our prices to gain our desired profit. The profit disappears, but the costs stay. For this reason they are called "fixed" cost. All fixed costs must be calculated before a profit can be declared or taxed. The name of this game is competition—something American workers are learning more about these days.

However, if we are able to produce a desirable product or service at a price that will compete favorably, then our possibilities for volume sales and expanding our business are enormous. Thus, the profits generated will far exceed any extra costs it might have been necessary to pay for higher wages and benefits in our earlier example. You have noticed that we did not say how much we were paying in wages and benefits, we only said that our product could perform "favorably" in competition. This is very important, as wages are the principle factor in the cost of manufacturing or servicing anything and they are fixed. Once set they are in place and are very difficult to reduce. Usually, rather than go through the pain of trying to convince his workers to take less, he simply lays them off.

If workers demand that all of their input—all three elements of their Capital Compound plus the synergism created—be compensated for in wages and benefits, they create a "fixed" cost. This fixed cost has to be passed on to the consumer, and remember "demand" is necessary before we can establish a true value of our product or expect success, in the marketplace.

To some extent we can create our own demand by using advertising. And, it's true that advertising will do wonders, but up to a point. Consumers have proven that their pocketbooks determine demand more than loyalty to brand names or product lines. Unless we are selling a luxury item that has demand because of its unique desirability for the wealthy (where price becomes less of a factor) then competitive cost to the consumer is going to be one of the important elements of our success.

LABORISM IN ACTION

As a result of all this, companies on the one hand are always trying to sell their employees on the fact of the company's need to remain competitive. On the other hand, labor is continually complaining that the company is making more money than they deserve to be making and that labor should be paid more. You can see that both are right and wrong at the same time.

Employees of the Chrysler Corporation reacted predictably to Chrysler's record profits in 1983. The employees, represented by the UAW, had agreed to accept wages lower than the industry average. This means that they were willing to help the financially ailing Chrysler Corporation improve their profitability by accepting less in wages and benefits than other UAW members working at competing companies.

Under the good leadership of Lee Iacocca, the corporation proceeded to produce a quality product at a competitive price—the result was record profits. It must be understood that the profits were made *after* the sales, not before. It's obvious that the leaders and members of the UAW don't understand what occurred, for as soon as the profits were made, the union representatives called a meeting and demanded that they be given their fair share of the profits. Great! But, in what form did they request their share be paid? Wages and benefits—what else?

But after all, what could we expect? After a hundred-odd years of insisting that the only way for labor to get a fair share is through higher wages and benefits, how could we expect a different response? They acted very predictably.

Now Chrysler has a "new" fixed cost on the product that the workers hope will give them a bright future. A future, by the way, that will regularly require more demands for more increases in "fixed" costs to satisfy an erroneous notion of wealth and how to increase it! It doesn't make a lot of sense, does it? Maybe we can get this chicken to lay two eggs a day instead of just one, what do you think?

What is the conclusion? Were the workers justly compensated with their lower wages? What about all of those profits? Don't the workers have a right to share in the successes?

THE INTANGIBLE ELEMENTS

The conclusion is this: the workers contributed four different kinds of input to Chrysler corporation. (1) They gave of their labor, (2) they gave of their creative initiative, (3) they gave of their spiritual inspiration, and (4) they contributed the value of synergism on their joint efforts. At the same time they took back from the corporation only wages and benefits for labor. All the rest stayed with the corporation, for now, and for as long as the corporation exists. The profits were made because the competition was met, and successfully challenged.

Contrary to popular opinion, the majority of the wealth was not created by the workers' labor, but by the other two thirds of the Capital Compound that are intangible. It was those intangible elements of the Capital Compound that made the volume of the product, the quality of the product, the beauty of the product, the marketing of the product, and the servicing of the product. In addition all this was enhanced by the added value of *synergism* that the corporate family created in their joint effort.

The workers' labor only gave physical direction to these intangible but *very real* elements of their Personal and Joint Capital Compound. Were it not for the workers' willingness to contribute those intangibles through faith in the company, the acceptance of a creatively structured labor contract, good workmanship, and team spirit, the whole thing would never have happened. Chrysler corporation would be history! It may yet prove to be history.

OVERVALUATION

Do the workers have a right to the fruits of their intangible contributions? The answer is both simple, and yet, somewhat complex. The direct answer is *yes*, but the more accurate answer is yes, *if they bargained for it!* In any case, it should be understood that a "yes" answer to the question does not mean that they should be given their share of the profits in wages and benefits. We can never forget that wages are a fixed cost and go directly to the cost of the product to the consumer.

It is always easier for us to believe more in things we can see than in things we cannot see. We feel a certain inadequacy about the aspects of our life that we can't see. We aren't sure of them; our senses don't as readily perceive them. As a consequence, this uncertainty increases our tendency to fear them. As a result of this, we have learned to put more confidence in the physical and less in the intangible. It is largely due to this fear of the intangible that the labor element of the Capital Compound is consistently overvalued.

Remember what we emphasized in Chapter Two when we focused on our right to access the equity productivity of our Personal and Joint Capital Compounds? It is important to review this one more time.

ACCESS TO EQUITY PRODUCTIVITY IS A GOD-GIVEN RIGHT

It is important to focus attention on the fact that Kelso and Adler did not mention anything concerning the intangible elements from which all technology is derived. This is simply because, for one reason or another, neither one had ever considered such technological developments as being associated with a gift from God. To Kelso and Adler, the evolution of technology and the need or advisability to share it with workers is simply a means to an end. Simple enough right?

No, wrong! There is a distinct and very important difference in recognizing the advisability of a new financial instrument, such as the ESOP for which Kelso is famous, and the understanding that the product of our Personal and Joint Capital Compound is the crucible of our inalienable right to "life, liberty, and the pursuit of happiness."

These we have as a gift from God, whether we recognize it or not. Anything less leaves open the door for others or governments to step in and say that we have no right to share in the equity creating portion of our Personal and Joint Capital Compounds. Without this right there is little or no substance to the hope offered in "life, liberty, and the pursuit of happiness." With no right or freedom to our portion of the wealth we have singularly or jointly created with others, we are left with only wages and therefore we will have allowed others to expropriate the productivity of our Personal and Joint Capital Compound. This is *not* what our founding fathers had in mind.

FEAR MOTIVATION

Fearing the two intangible elements of our Capital Compound, we find ourselves in a similar predicament to Job, an ancient Hebrew man of great wealth who lost all he owned. He gives us a clue to why he may have lost it when he exclaims, in the book of Job 3:25-26, "What I feared has come upon me; what I dreaded has happened to me, I have no peace, no quietness; I have no rest, but only turmoil."

Insistence on the importance of the physical element of our Capital Compound is what has sidetracked leaders and men of good intentions for centuries. While great attention was being given to the *labor* element, the rapid and increasingly dynamic contribution to the production process of the two intangible elements—*creative initiative* and *spiritual inspiration,* illustrated in the explosive nature of technology—was being ignored.

Not until the dawning of the Industrial Revolution, which began the process of duplicating man's muscle and mental processes with the product of the intangible elements, did we really begin to take notice! Unfortunately, instead of embracing the intangibles, labor ran from them. No one sounded the alarm of fear more eloquently or ran faster than did Karl Marx and Friedrich Engels.

A FORM OF CHARITY

The adjective which best describes our primary means of distribution is "laboristic." As a result of our extreme orientation toward "laborism," we have created an economy that has within it the seeds of instability continually going from depression to inflation and back again.

Laborism as a primary method of distribution with "full employment" as a national goal has many disadvantages. It insists on a type of charity. In essence we are saying that in order for the workers to be able to buy what the factories and service organizations have to sell, they must be paid more than their labor element is worth. The portion that is overpaid is a type of charity or welfare that is built into the system. Remember, we are only talking about the physical labor element and how that it is not worth as much as the two intangible elements, creative initiative and spiritual inspiration.

What expensive charity and welfare! This overpayment has to be added to the selling price of every product and service in the country. It becomes a part of the fixed cost. All of our efforts to compete have that extra load tagged on, both in our own domestic markets and in the international markets. It's no wonder we've got problems.

The only true beneficiaries of this "charity" are not the workers or the unions who support it, but the union bosses and the politicians who support them! Follow the money.

THE FRUITS OF LABORISM

Now, with this new fixed cost, we all need to be paid more money than our *labor* element is worth. *Inflation* is the result. Have you ever heard of "Buy now and pay later?" Well, welcome to pay later! No one in the history of the world has practiced this misguided assumption more than Barrack Obama. The problem is that we, the taxpayers, are left with the bill. Only the Progressive Socialists can enjoy that kind of change!

When inflation as a result of the overvaluation of labor has put us in a noncompetitive position, layoffs and unemployment are inevitable.

The phenomenon of "stagflation" is a unique factor of our modern, technologically-oriented economy. When stagnation is evident in some sectors of the economy while a boom is occurring in other sectors, it is largely a direct result of overvalued labor. Those areas suffering stagnation are on the trailing edge of technology, usually due to too little investment in information technology and research and development, in addition to the overvalued labor costs. The result is a falling behind in competition and loss of their market share, which forces unemployment and layoffs in that sector.

Laborism's impact on the unemployment and welfare sectors of our economy is grudgingly accepted by society. It's like playing the game Hot Potato: everyone keeps it moving from one administration to another, but none knows what to do to stop the game. The "hot potato" they keep passing around is only a symptom of the cause. The cause or origin of the problem is rooted in that early decision to accept labor as the only source of wealth, and wages as the primary method of distribution. It automatically follows that Full Employment becomes a national goal in order to provide the wage opportunity that has been accepted as the primary method of distribution. I just love that Jobs Bill, don't you?

The cancerous nature of this belief and practice is evidenced in what we see happening now with regards to the Obama administration. Obama has now agreed, along with his big labor bosses in the AFL-CIO, to bail out the largest corporations who use union labor. He and his advisors do this under the guise that these companies are too big to fail. I wonder in what smoke-filled backroom union hall that was decided.

What could possibly be better than Full Employment? How about receiving an income from ownership in the "instruments" of your Capital Compound, and the synergism that you contribute to the Joint Capital Compound group effort? As your percentage of such ownership increased, your dependency on a wage and other types of benefits would decrease. You might say you could even become "financially independent." Not independent of others, but certainly independent of toil.

Ownership of our total Personal and Joint Capital Compound—and just distribution of its contribution to the total group effort—would eventually reduce unemployment to zero, except for those unfortunate few who actually cannot do anything. In reality these are very few indeed.

On the other hand, consumerism for the sake of keeping factories and service organizations operating will always tend to increase unemployment. This appears contrary, even backward, but only until one takes into account the dynamics of the two intangible elements of the Capital Compound. It is from these elements that we derive all of our technological advances and we have already seen the tremendous impact and power each new discovery brings to the production process.

Consumerism in an economy based on the Capital Compound Theory of Value would of course certainly be robust, but without being contrived or stimulated by deficit spending. Individuals who have ownership in the capital elements of production—i.e. the results of our intangible elements— would not require any government, state or federal, to pass a deficit spending bill to stimulate consumer spending. These individuals have their own income flow separate from any wage or salary and are perfectly capable to use cash or to establish a line of credit when they see a need to do so. Furthermore, the personal financial management actions of such individuals would not create any inflationary actions.

Welfare shares equally in its relationship to our insistence on a laboristic form of distribution of wealth. Having failed to provide the individual wage earner a place of ownership in the capital instruments of our Joint Capital Compounds, we have created a great many more welfare recipients than necessary. In addition to that, we make those on welfare a part of the consumerism cycle by forcing them to identify with only a wage-like payment, a monthly stipend from the government. Perhaps the prospects of ownership in capital instruments would even light a spark in the dulled eyes of some of those forgotten souls.

A LOGICAL RESPONSE

Let us at last address the question asked earlier in the chapter, namely, how can we effect the necessary changes in America to give us a more equitable distribution of wealth than we now have, while at the same time offering some hope to those millions of the world who are plagued with the disease of poverty?

It would seem that the answer is now clear. The Capital Compound works and we have no need to fear the intangible elements of its composition. Our focus on the distribution of wealth must broaden to include the fruits of the intangible elements of the Capital Compound. Creative methods of promoting ownership for everyone in the instruments of our Joint Capital Compound are mandatory if we are ever to break the bonds of laborism. When we are successful in the refocusing of our distribution emphasis for Americans—and the way we conduct ourselves relative to our nation's businesses, corporations, government and institutions—we will set a new stage for a new era! By the grace of God, we will not only drastically improve our own lives, but we will establish real hope for our fellow brothers and sisters caught in the birth pains of emerging cultures and societies.

Alexis de Tocqueville, a French statesman and political philosopher, became known for his book *Democracy in America,* published in 1835. He wrote it after a visit to the United States in 1831. With near prophetic vision he foresaw many of the social, political, economic, and cultural problems that we would confront as a society developing under what he called "conditions of equality."

De Tocqueville looked to the future with faith that Providence would decree and guide the progress of society from conditions favorable to the few to conditions favorable for all. He both challenges us and warns us to solve the problems incidental to such progress. His closing remarks were these words:

> The nations of our time cannot prevent the conditions of men from becoming equal; but it depends upon themselves whether the principle of equality is to lead them to servitude or freedom, to knowledge or barbarism, to prosperity or wretchedness.

Let's say that again!

> *But it depends upon themselves whether the principle of equality is to lead them to servitude or freedom, to knowledge or barbarism, to prosperity or wretchedness.* (Emphasis mine)

The following is a Tea Party announcement made after Speaker of the House Nancy Pelosi and her minions forced the Health Care bill down the American people's throat, on March 21, 2010:

> Arm twisting, threats, buying votes and political dirty tricks are now the normal operating procedure for the new regime …

The Obama putsch has taken roots and is thriving in the new era of the expanded welfare state. The American people said no, Conservatives said no, but it makes no difference when the oligarchy says "Yes we can" in spite of the will of the people . . .

If we stop the madness and restore America now we can halt the bleeding of our freedom. However, if we don't, they will come for our sovereignty and our beloved nation will be flooded with millions of illegal aliens further converting America into a Socialist state. They will come for our guns, and our gun rights will fall into their hands. They will come for more taxes, collection of healthcare premiums, along with penalties and fees will be heaped upon the American public. A new form of debtor's prison will emerge, not of walls, bars and cells, but the loss of rights. Privileges will be rewarded to those who obey while the disobedient are punished by privilege banishment, they will come for our money. The thrill of risk and reward will vanish in a cloud of the greatest good, and free enterprise will become the unwilling slave of the new enlightened order . . .

To restore our beloved nation from the pestilence of National Socialism and liberal extremism we must rise up, pledge our life, liberty, and sacred honor. We must vote them out! We must file lawsuits to stop egregious legislation. We must march in protest. Most of all, we must build our ranks, recruit, grow, and train our members on how to restore our nation. We must not be afraid of this task, for this is our land, our nation and our heritage, freedom is our national treasure. Our freedom, liberty, and self governance have fallen upon our shoulders and it will be our effort that will stop the greatest runway government in history.

SUMMARY

We have reviewed the controversy over the distribution of wealth. We know that no matter how much we invest our Personal and Joint

Capital Compounds into the process of productivity, we can never be equitably compensated through an hourly wage or set salary. Leadership of industry the world over has consistently been led astray through the misunderstanding of these powerful compounds.

It must be clear to us now that we cannot expect the current leaders of industry or anyone else to invite us into the halls of equitable distribution of our justly earned wealth. We the citizens, the workforce of America, must take that responsibility as our own. Only through the consistent and persuasive presentation of the facts of the Capital Compound Theory of Value will we get the attention of those we serve. We cannot keep silent.

They are more ready to listen and to understand than we give them credit for. Our success is also the crucible of their success. Why would they want to remain in the dark on such a dynamic issue as this? The truth is that they don't. It is only our fear and lack of confidence in the reality of our Personal and Joint Capital Compounds that will hold us back. Remember, we have this gift of creating wealth from God, not from other men. We must build again what our colonist forefathers experienced. We all must join in the chorus of "Arise America, rebuild your God-given Capitalist foundations!"

Chapter Six

Understanding Obama and
Sharing the Wealth

There is an old saying that, although a little crass, nevertheless illustrates a vital point that we who love the heritage our forefathers have left us in this great Republic must consider: "Keep your friends close and your enemies closer."

When we know that there are major and powerful enemies lurking inside of our beloved America, we must know them, understand them, and keep them close so as to never be taken by surprise.

UNDERSTANDING THE MIND OF OBAMA

Obama did not just wake up one day and decide to be a Progressive Socialist. From his earliest days living in Somalia, he was surrounded by individuals from within and without his family who believed in the doctrines of Marx and Engels. He was taught to respect their theories of Communism and sharing the wealth. He continued these studies after coming to live in the United States. He developed many relationships with others who believed the same as he. He worked with and assisted organizations such as Acorn. He became close with many extremists, some who even escaped conviction and went on to hold high positions in major universities as Bill Ayers did.

It is extremely important that we understand Obama's perspective of truth and common sense in order to see clearly what he is and what he is all about. Ideas have consequences and Obama is the consequence of the Progressive Socialist concepts that he has studied and embraced. There are obviously some who agree with him and who applaud him as a great thinker and a great leader. This is nothing new, as the world has had many despots who were held up as the greatest and the best ever. A person's reputation depends upon who you are speaking with. The following is an in-depth look at what President Obama learned and embraced during his formative years. The ideals and principles of Marx and Engels presented here are the ideas and principles that drive him and keep him focused.

His unfaltering faith in their truth is what motivates him to repeatedly declare his conviction that he will succeed no matter what. It is impossible to ignore his extreme focus on what he is doing and, despite setbacks from Republicans or those of his own party, he always ducks and weaves and comes back to the same point with a different angle. The only time that Obama will cease to be a threat to America is when he is out of all political power.

BALANCE NEEDED

There is little, if anything, that can so easily touch our senses as satisfaction from receiving what we would consider to be a fair compensation or, on the other hand, to believe that we somehow have been cheated. The reactions are, of course, opposites, and usually more intense in the latter situation.

MAIN ISSUES

Before we draw conclusions as to who should get what as we distribute the wealth generated by our Joint Capital Compound, several issues need to be considered.

1. If labor is not the total element of exchange for which we should be compensated, and if this element is overvalued driving up inflation, what other assets of our Joint Capital Compound are there to which we can lay claim?

2. If there are other assets to which we can lay claim, then what rules of ethics and justice must we apply to affect such a claim?

3. Assuming that some additional compensation is forthcoming, what forms should it take, if wages are not the answer?

WHAT OTHER ASSETS

Despite the fact that a long list of able economists has adequately shown the Labor Theory of Value to be an incorrect premise, they have not silenced the cries of the world's masses for justice. There's a motivation behind the premise that keeps selling the Marxist philosophy to the poverty-ridden multitudes of the Third World countries. Any attempt to further disprove the arguments of Marx relative to his Labor Theory of Value is pointless—unless, that is, a proper understanding of his motivations lead us to a superior solution. As we examine Marx's motivation for his theories, we will come to see very clearly the new wealth to which we can lay claim.

The basic human weakness to be attracted to that which seems to be free is the crucible where Progressive Socialism attacks all humanity. Americans are not immune to this evil virus. Yes, it is easier to spread within the Third World countries as we have said, but look at what is happening to our own beloved America even as I write these words. Fortunately, the majority of us have not yet fallen for the lies and deceitfulness of Progressive Socialism, but millions of us have. May God bless us with a true awakening, not just for us who know the truth, but for the nation as a whole!

EARLY CONDITIONS

In order to more fully appreciate Marx's Labor Theory of Value, let's once again recall to mind the setting in which it was originally formed:

These difficult and inhumane environments of the early exploitations of the worker aroused in the fertile minds of Marx and Engels a desire to know the cause and, if possible, to discover the solution to their plight. From the perspective of those who suffer poverty today, both in America and in the emerging nations of the world, the plight still exists. Consequently many of the same conditions of Marx's era exist today for such people, as does the desperation that goes along with it.

After a long study of the problem, the conclusions that led Marx and Engels to the Labor Theory of Value were centralized around two discoveries which they believed they had made.

The first proposed discovery was the Value of Labor Power. Very briefly, it says that when a worker performs physical work without the help of a machine and uses that labor power to cultivate or produce a product from any raw material or natural resource, that it will take one hundred percent of the labor power effort to supply the worker's basic necessities.

The second proposed discovery was the Theory of Surplus Value. Also very briefly, it says that when a worker's environment is invaded by a machine, his efficiency goes up and he is able to earn (or "produce" as it were) his basic necessities with less than the one hundred percent use of his labor power. This assumes that in both instances the worker was working the same number of hours at the same effort. Therefore, the conclusion is that after the point of having produced enough to provide the worker's basic necessities, any extra hours spent working in the factory or on the land were hours that the "Capitalist" who owns the factory or the land had expropriated from the worker.

SOMETHING FOR NOTHING

In modern America, with our current postmodern approach to laborism, this seems absurd. Well, except for the Progressive Socialist elitists, who are smarter than all of the rest of us. We would assume that the money earned in those extra hours was just compensation for the extra work accomplished, right? Yet, in Marx's day, that was the very point.

A worker was paid by the day, not the hour. Workers were customarily paid a standardized day-wage which kept them on the verge of abject poverty and misery, regardless of any increase in their efficiency. When increases in pay were forthcoming, it often meant an increase in the number of hours that one had to work to receive it, thus the advent of the twelve and fourteen-hour workday.

Because of that situation it is easy to understand the importance of organized representation of labor to the workers of that era. Gradually the movement gained enough strength to force the shortening of the workday. Each victory was difficult to gain, and often ground was lost meanwhile in some other area.

What Marx and Engels were able to realize and expose was that the owners of capital instruments (factories, machines, land, and so on) were gaining something for which they did not have to pay. This gain was created in the form of what Marx called the surplus value, that value gained in the extra hours the worker produced for the benefit of the project beyond covering his basic necessities. The problem to Marx was simple. The advent of factories and increasingly more efficient machinery meant that the worker was becoming ever more efficient, and was therefore providing ever more *unjust* profits for the owners of the capital instruments.

Can't you just hear Obama and his ilk complaining about ***unjust profits?***

MARX PREDICTS THE FALL OF CAPITALISM

The next conclusion that Marx drew was intimately related to these two proposed discoveries. He used this information to formulate his Rate of

Surplus Value theory. The theory states that since the only real source of wealth is labor, and the introduction of more and more machinery only serves to make labor more efficient with the same wage, the "surplus value" is forced to increase.

However, he noted that in each case studied it also meant a lowering of the unit price for each product produced. Mass production and ever increasing efficiency made for less expensive consumer items then, even as it does now. As the increase in more efficient machinery and production methods continued, it meant fewer workers were needed to accomplish the same amount of production. But with fewer workers and a higher investment in machinery, the "surplus value" which had made the factory owners rich would also decrease, aided by a further drop in value of the product. As a consequence, the future of Capitalism as Marx and Engels knew it was, in their opinion, doomed.

Their conclusion was that labor and capital could not survive together. This set the stage for the argument for total expropriation of all capital from private hands. This is the same conclusion that exists today in the minds of every Progressive Socialist, and none are more committed to its fulfillment than Barack Hussein Obama! Watch for more expropriations!

Marx said in his Theory of Surplus Labor Value,

> This mode of production produces a progressive decrease of the variable capital [physical labor], as compared to the constant capital [factories and machinery], and consequently a continuously rising organic composition of the total capital. The immediate result of this is that the rate of surplus-value, at the same degree of labor-exploitation, expresses itself in a continually falling average rate of profit. This progressive tendency of the average rate of profit to fall is, therefore, but a peculiar expression of capitalist production for the fact that the social productivity of labor is progressively increasing. This is not saying that the rate of profit may not fall temporarily for other reasons. But it demonstrates at least that it is the nature of the capitalist mode of production, and

a logical necessity of its development, to give expression to the average rate of surplus-value by a falling rate of average profit. Since the mass of the employed living labor is continually on the decline compared to the mass of materialized labor incorporated in productively consumed means of production, it follows that that portion of living labor which is unpaid and represents surplus-value must also be continually on the decrease compared to the volume and value of the invested total capital. Seeing that the proportion of the mass of surplus-value to the value of the invested total capital forms the rate of profit, this rate must fall continuously.

Though most of us find Marx not too easy to read, he is worth the effort. Marx and Engels were both convinced that applied technology in the form of machines and automation would ultimately reduce the "surplus labor value" (or profits) to such a small amount that the whole capitalistic industrial system would go bankrupt. They measured the difference between the cost to produce an item and the sale price, and called it the "rate of profit."

Their solution, based on what they felt to be undeniable universal laws of dialectic materialism, was simply to take over and share the surplus value between the newly liberated workers and the new state—no halfway adjustments would suffice!

Amazing as it may seem to all who have not accepted these theories as fact, our beloved America is now in the grip of a president who believes these theories as fact. More importantly he believes the solution is total centralized control. He hasn't said that in so many words, but his stated goals must ultimately include total centralized control in order to be realized.

LABORISM ADDS TEMPORARY NEW LIFE

The reason the Western noncommunist economies continue to thrive despite Marx's findings is due to the freedom to own property and the rise of more equitable treatment of labor without organized unions. Without the ownership of private property, better labor relations, and the growing realization on the part of manufacturers that a poor populace made for a poor marketplace, there is little doubt that world financial collapse would have resulted as Marx predicted. The system as practiced has, of course, suffered and continues to suffer from different problems than those that were predicted. Yet, imperfect as it is, the free marketplace, good labor relations and the opportunity to own private property are the conclusion of what our Joint Capital Compound success has given us at this point in time and history. Our current status seems to be very successful to many, especially those who aren't paying attention and have allowed themselves to become politically and financially illiterate.

The truth is something quite different. Our American financial prosperity, as well as our individual freedoms, are so close to being lost that only a grassroots movement of historical proportions will successfully bring us back from the brink to a solid foundation of true principles and character.

TECHNOLOGY FORCES THE ISSUE

One of today's major challenges is that the dynamic productivity of the intangible elements of our Joint Capital Compound (as expressed in technology) has outweighed our wildest dreams. On the other hand, private capital invested in the instruments of production and service is so concentrated in the hands of a few that, if there are no changes, the absolute control of the worker is inevitable. Only if major efforts are adopted to distribute ownership of capital instruments to all do we have hope for continued prosperity and the preservation of individual freedom such as most have enjoyed up until now. If we take such action, even greater things may be in our future.

The Progressive Socialist ideology adopted by many countries of the world makes them, in many cases, no more than giant corporations owned by the government and staffed by the populace, who neither vote policy nor bargain for wages. They are *national monopolies* that increasingly control a larger share of the world market, not the least of which are underdeveloped countries. Saudi Arabia, Iran, Syria, Lebanon, and China come to mind, plus many others of less notable distinction.

SUMMARY

It is said that hindsight is 20/20, but just because we can see much more of the iceberg nearly 150 years later, doesn't detract from the fact that Marx and Engels realized something of value was being received for which no payment was made. A gross expropriation was in the process and they honestly thought they had discovered the solution. The technological breakthroughs of the Industrial Revolution forced men such as Marx and Engels to consider the intangible elements of the Capital Compound. But from their myopic, faithless, dialectic point of view—being able only to perceive a minuscule part of the tip of a giant iceberg—they were not able to accept the intangible elements as a blessing, but rather saw them as a curse to the worker.

If they and others of that day—such as statesmen, union leaders, and organizers— could have had our vantage point, what steps would they have taken? What steps would such men advise today, if they were alive to see the exploitation of high technology reaching to the point of Artificial Intelligence? Would their answer be job security clauses in wage contracts and expropriation of assets, or would they perceive a need for distribution of ownership in the capital instruments of production? We might easily fool ourselves into believing that they would do differently. However, we are the future and we have the opportunity for a more perfect perception. We must ask ourselves: What will we advise and what will we do now with what we know?

Obama and the elitist ilk that surround him are only the tip of the Progressive Socialist iceberg. Fortunately, we do know how they think and why they think what they think. This is a truth that will set us free! We do have faith but we need to pray for more. We do have great valor and courage but we need to pray for more. We can and we will act on this truth: *Keep your friends close and your enemies closer.*

We know who the major and powerful enemies lurking inside our beloved America are and we do know them for who they are. We understand them and are keeping them close so as to *never* be taken by surprise.

Chapter Seven

Refocusing the Facts

From ancient times there are recorded instances of efforts by workers to bargain for better pay for their services. Unfortunately for them and for all who have come afterwards, everyone down to the last person always bargained for cash and cash only. This is evidence of the earliest beliefs by mankind which carries down to our present day dilemma over equitable distribution of wealth. The advancement of technology, the fruit of our intangible elements in the Capital Compound are now forcing us to look closer at who we are and how really do we participate in the creation of wealth.

EVERY WORKER MAKES A CONTRIBUTION

We have established that the intangible elements of the Capital Compound, creative initiative and spiritual inspiration, are the source and the driving power of all physical labor and the producers of all our technology. Some individuals, we will find, will be more justly compensated than others for their part in the development and management of technologies. Yet it cannot be denied that without a team effort within a company, or a team effort within a society, the maximum potential wealth to be gained by such technology and its application will not be realized. It is only reasonable to conclude that all workers in any industry have made a contribution to the whole, and cannot ultimately be denied just compensation. Obviously, since an increase in wages is counterproductive to competitive profits, such compensation must be in some other form.

Realistic examination of the workplace reveals that we still have most of our old problems; they have just been dressed up differently. We need a new perspective and a new direction if we are going to truly overcome these handicaps. Not until we, as a people, are willing to embrace the reality of the intangible elements of our Capital Compound, bringing that acceptance into focus and balance with our understanding of the labor element, will we have the power to overcome our inequities of distribution! Needless to say, if we are unwilling or unable to face and conquer our own inequities, how can we possibly hope to offer more than slogans to the oppressed of the emerging nations?

ETHICS AND JUSTICE

Having established that injustice in the distribution of wealth is a valid claim, what rules of ethics and justice must we apply in restructuring the distribution process?

Our Declaration of Independence lays down the basic foundation for the measure of justice that we must apply in any distribution process.

> We hold these truths to be self-evident, that all men are created equal, that they are endowed by their Creator with certain unalienable rights that among these are Life, Liberty and the pursuit of Happiness."

The four basic areas of justice our constitution challenges us to address are:

1. Our equality in creation

2. Our right to Life

3. Our right to Liberty

4. Our right to pursue Happiness

Our equality in creation is relative to our distinction from animals and other created beings. We are all alike in our natural possession of the dignity of being human, and as persons having the natural endowments of reason and freedom of choice which confer on all the capacity for active participation in economic, political, and religious life.

Our right to life is the most fundamental of all our rights. The right to existence is basic. However, much more than the right not to be murdered or maimed is implied. A person cannot live long without some reliable means of subsistence. Therefore our right to life includes the right to be able to acquire a lawful means of subsistence. By extension, this also includes private property such as land, houses, personal possessions, and capital instruments for use in production.

Our right to liberty is a further extension of our right to existence. If one is held in chattel slavery by others, or so restricted by oppressive laws as to make it impossible to acquire a lawful means of subsistence, then there is no meaning to "liberty" and "life." Freedom of movement is also implied, as is freedom of personal expression in all areas of our cultures and society.

Our right to pursue happiness encompasses the first three and further extends our limits. However, to pursue happiness, and to allow all men and women to do likewise, most assuredly requires that my pursuit of happiness does not in any way restrict or deny your pursuit of happiness and vice versa. The right to pursue our individual happiness includes our pursuit of economic stability and security. This basic right also encompasses our political freedom to participate in all forms of government and institutions which regulate our lives, and our religious expression of assembly and worship.

Our equality in creation is vindicated when we remove all obstacles from the path of equal opportunity for all, regardless of sex, creed, color, or race. The only expression of inequality should be in our acquired abilities, not in the opportunity to acquire them. Only our own natural limitations should dictate the limit to which we can add to our acquired abilities and grow.

Our right to life began to be vindicated during the early period of the Industrial Revolution. As a direct result of the efforts of such men as Marx, Engels, Smith, Ricardo, and others, workers' demands began to at least get some attention. Their right to life was finally being recognized. This right undeniably exceeds the right to a bare existence, and includes a right to even the best rewards: "... not only in the meanest and coarsest, but likewise the richest and choicest of them all . . ." was the way the Mechanics Union put it.

Unfortunately, history shows that all concerned have consistently adopted the same basic faulty plans in an attempt to bring justice to the worker. Economically speaking it is impossible to distribute a profit before it is made, or to distribute a portion of ownership in new growth before such growth actually takes place. The "original sin" so to speak, was perpetrated from the very beginning of mankind's early experiences when it was assumed that all a person gave or offered to an employer was physical labor.

A BARGAIN IS A BARGAIN

Consequently a bargain was struck. This was a bargain that accepted distribution through labor, as the sole instrument for compensation. Ever since then, restitution for acts of expropriation committed against the workers of the world has, for all practical purposes, been attempted only through wages and benefits, rather than joint ownership. The sentimentality for this type of bargain can be appreciated, but not the disastrous results.

In order to diminish the temptation to incorrectly place all of the blame for these unfortunate developments on only one side of the relationship, we need to consider the fact that organized labor initiated the argument in favor of labor as the only source of wealth and wages, as the proper method of distribution. Even though owners and managers of capital instruments of production did not offer any creative alternatives, the fact remains that labor successfully "bargained" for what they wanted. It stands to reason that if they had been sensitive to the value of what they were sacrificing, they would have bargained for joint ownership compensation from their Capital Compound contribution, and would have won.

In Matthew 20:1-15 Jesus told his disciples a parable that illustrates a similar situation.

> For the kingdom of heaven is like a landowner who went out early in the morning to hire men to work in his vineyard. He agreed to pay them a denarius [a Roman silver coin called a Penny] for the day and sent them into his vineyard.
>
> About the third hour he went out and saw others standing in the marketplace doing nothing. He told them, "You also go and work in my vineyard, and I will pay you whatever is right." So they went.
>
> He went out again about the sixth hour and the ninth hour and did the same thing. About the eleventh hour, he went out and found still others standing around. He asked them, "Why have you been standing here all day long doing nothing?"
>
> "Because no one has hired us," they answered.
>
> He said to them, "You also go and work in my vineyard."
>
> When evening came, the owner of the vineyard said to his foreman, "Call the workers and pay them their wages, beginning with the last ones hired and going on to the first."
>
> The workers who were hired about the eleventh hour came and each received a denarius [silver penny]. So when those came who were hired first, they expected to receive more.

But each one of them also received a denarius. When they received it, they began to grumble against the landowner. "These men who were hired last worked only one hour," they said, "and you have made them equal to us who have borne the burden of the work and the heat of the day."

But he answered one of them, "Friend, I am not being unfair to you. Didn't you agree to work for a denarius? Take your pay and go, I want to give the man who was hired last the same as I gave you. Don't I have the right to do what I want with my own money? Or are you envious because I am generous?"

LABOR BARGAINED FOR A PENNY

The only workers in the parable that "bargained for a penny" were the first workers. All of the others left it up to the lord of the vineyard. We see that contracts and agreements, and bargaining for an equitable exchange, date from man's earliest recorded history. In the case of the early efforts of organized labor to get a fair distribution of the wealth being generated, the wrong bargain was made. They bargained for the penny, while they could have bargained for part of the profits from the vineyard. Further efforts by subsequent generations of organized workers only exaggerated the error, and have compounded it into our present day situation.

This is not to suggest that, had the workers left the matter up to the owners of factories and instruments of the Joint Capital Compound, they would have acted with the same goodness as the lord of the vineyard in the parable. Agreements and bargaining are certainly needed. The tragedy for all concerned is that vision and insight were lacking and a "wage" was the only method bargained for in the process of distribution.

Fast forward two thousand years to our current situation and we see how far the arrogance of man can reach. We now have a new savior, a new "messiah," a new anointed one, who claims to be the new lord of the vineyard. He is all-wise and all-knowing and he alone can rightly share the wealth of America, not to each according to their abilities, but to each according to their needs.

EVERYONE LOSES

This unwise choice has led to losses on both sides of the relationship. In efforts to compensate more fairly, companies have been "bargained" into positions of paying more for the labor element than it is worth. This constitutes an effort to pay out a portion of anticipated profits before such profits are made, forcing an increase in the fixed cost, which ultimately creates losses instead of profits. The foreseeable loss of profits is countered by creating further demand for labor-replacing technology.

Furthermore, workers who attempt to exercise their right to meaningful subsistence at the expense of the owners of the capital instruments for whom they work perpetrate an injustice equal to the one that they have suffered. In so doing, they feed the destructive process: two wrongs don't make a right.

Another very important area of unjust treatment toward the owners of companies and the instruments of production strikes at the very heart of one of our most basic liberties, the right to own private property. This right does not limit property owners to the joy of a mere piece of paper that declares their ownership, but implies that they are entitled to receive the profits that the property produces for them.

Imagine if you can a state governor assuming the right, under new laws, to hold the power to declare each year whether or not apartment owners and commercial real estate owners could collect the profits from their rents or not. Such actions would probably lead to armed revolt, or at least public outcry. Actually this process is undertaken every year in the form of property taxes which must be paid whether there is an actual net profit or not. The state, then, does have its creative ways to expropriate our wealth.

Now, with the Progressive Socialists in power in America we can expect that new creative ways to deny rental property owners their just profits from the rents they charge on their properties.

One would not think it possible that such a situation could ever exist in America. Not only does the situation exist in the form of property taxes, it also exists in the form of indirect expropriation through corporate taxes. We have been so brainwashed by the Progressive Socialists that we have come to assume such things to be correct! What we are referring to in the latter is the act of expropriation of profits from corporate stockholders in America. Each corporation, under our present corporate law, is allowed to have its board of directors decide each year if they are going to declare a dividend. It would almost seem that they call it a "dividend" so as to make it less obvious that what is really being referred to: a profit.

If a corporation chooses to declare a dividend, then it must first declare a corporate profit on which it must pay a corporate income tax. In order to avoid this tax, a common practice is not to declare a profit, but to plow back as much of the gross revenues as possible into expanded facilities and equipment. The standard argument in defense of such actions is that the expansion increases the value of the company and therefore increases the value of the shareholders' stock. Reality dictates otherwise, as it is impossible for the shareholders to be assured that they can immediately take their shares to the marketplace and realize a proportionate increase in price to match the company's "retained earnings."

The laws we now have and enforce for control of corporate profits is demonstrative of our insistence on laborism as our primary means of distribution.

Inequities of this type will have to be resolved before we can assure justice to all property owners and allow for proper distribution of wealth through shares in capital instruments.

BOTH WAGES AND EQUITIES NEEDED

A distribution process that includes both wages (cash payments) and equities (shares in the company) is the only possible fair method that does not violate the rights of any of the parties involved in the production of wealth, and that has the *eternal qualities* of our intangible elements. The distinction we now make between management and labor—between the assumed owners of the instruments of our Joint Capital Compound and those who are supposedly only involved in a labor element process—will have to be changed. All must become *proportional owners* of the entire entity, and share in the efforts to protect its growth, health, and assets for the benefit of all, not for just the few.

Henry Ford provided the first example of the industrial establishment supporting consumerism by announcing the first five dollar workday.

The consumerism concepts which followed, those of politicians, labor leaders, and economists, have served only to further the incorrect assumptions underlying "laborism." If sharing profits is possible through the wrong methods (and sometimes for the wrong reasons), it is not difficult to believe that sharing in ownership by all, in the instruments of production of our Joint Capital Compound, through both correct and "right" methods is both possible and practical.

If only one half of the relationship were to benefit from such changes in our basic attitudes toward distribution, then serious doubt could be raised. But when both sides stand to gain, only ignorance, greed, and pride could allow the continuation of our present direction. The ultimate losses of personal freedom, justice, and liberties will affect all but the very few who will remain in total control. Ultimately even these industrial elites would lose when they are consumed by the very centralized power that supported them while maneuvering for total control.

In order to avoid such future disasters, every company and corporate leader and every worker in America must come to the realization that only equitable joint ownership of all capital instruments will ensure future financial security, which will not only continue our freedoms of private property, politics, and religion, but it will expand them to new horizons never seen before.

FORMS OF COMPENSATION OTHER THAN WAGES

Focusing on the Capital Compound from a perspective other than the labor element offers a refreshing view. Immediately, the doors of opportunity open onto a wide array of possibilities. It is not our intention here to present an exhaustive list of these possibilities, but rather to merely point our investigation in a creative and imaginative direction.

Considering our basic "laboristic" mentality, it will be more difficult for some than for others. But with a little imagination it is not hard to see the many different ways that one can actually contribute to the total intangible force of a company's Joint Capital Compound.

CONTRIBUTING WITH THE INTANGIBLE ELEMENTS

Many American workers would probably have difficulty believing that they contribute anything from their intangible elements. Let's review some of the different ways that workers can and do contribute to the total intangible force of a company's Joint Capital Compound.

It is easy to recognize the creative efforts of the engineer, computer scientist or the chemist, but what about the average manager or skilled worker?

The "hot spots" of our modern business environment are those immediate areas where all of the apparent action is taking place. These would be the laboratory, the engineering department, the planning department, and all of those other high-level centers of core activity. What all too often is overlooked by both management and the technical people involved is that all of this hot spot activity can be wasted at great loss if the total team isn't functioning well.

The connecting links between these hot spots are the infrastructure of office workers and skilled workers whose duties are to coordinate the multiplicity of directives needed to make it all come together. Even the janitorial department cannot be taken for granted. To accomplish the overall company efficiency and success requires good *internal* relationships within the company family and good *external* relationships with suppliers and customers of the company family.

Follow-through is the name of the game. No matter how brilliant the basic technological discoveries might be from any given scientific discipline, the execution of the transition from bare concept to the practical takes a high level of teamwork. Perhaps, when there are a great many "smart computers" utilizing artificial intelligence, such exacting group coordination skills will not be so important. (When that time does come, be sure you own a piece of the AI!) In the meantime the "human element" prevails and needs to be nurtured and appreciated much more than it is.

PERSISTENCE IS OMNIPOTENT

A good illustration of one of the intangible elements at work is persistence. You can't touch it, you can't feel it, you can't see it, you can't taste it, nor can you hear it. It is not perceived by our five senses, yet is very real. In Thomas J. Peters and Robert H. Waterman Jr.'s book, *In Search of Excellence,* what they call "skunk works of industry" (small units of unbudgeted research groups) thrive on persistence. Peters and Waterman say a greater percentage of major breakthroughs in technology come as a direct result of this one intangible ingredient more than any other single thing!

An unknown author has left us all a small gem of wisdom in the following text:

> Press on, nothing in the world can take the place of persistence. Talent will not: nothing is more common than unsuccessful men with talent. Genius will not: unrewarded genius is almost a proverb. Education will not: the world

is full of educated derelicts. *Persistence and determination alone are omnipotent!* (Emphasis mine)

At every level of human endeavor, one can find a use for persistence. No task is so small or relationship so fleeting that an attitude of persistence will not make a difference in the ultimate outcome or performance.

INTELLIGENCE AS A FORM OF VALUE

Building an information block relative to one's profession or desired vocation and staying on the leading edge of the direction of one's industry will serve every American worker well. Add to that sensitivity to developing one's own good self-image and people skills and you will find yourself in constantly increasing demand.

All signs indicate that we are at the mature stage of a great information gathering and distribution era. Twenty-eight years ago we were just at the beginning of this era. John Naisbitt saw early signs of this. In his 1982 book, *Megatrends,* he writes:

> In an information economy, value is increased, not by labor, but by knowledge. Marx's 'labor theory of value,' born at the beginning of the industrial economy, must be replaced with a new *knowledge theory of value.* In an information society, value is increased by knowledge, a different kind of labor than Marx had in mind.

Is this a true prediction? Yes and no. We can see now, from the advantage of twenty-eight more years of scientific discoveries, the predominance of "knowledge through information." However, Naisbitt's conclusion relevant to his suggestion of replacing Marx's "labor theory of value" with his new "knowledge theory of value" is myopic and fails to take into consideration the *distribution* of the wealth thus created. Who gets what and follow the money, are still the crucibles of success or failure in any equitable economic system.

In the case of Obama and his Progressive Socialist advisors, they simply say: "Leave the distribution and the sharing of the new wealth to us, we know best who should be compensated and how much." Their arrogance is so huge that they have already dictated the amount of compensation that can be paid to top management and how much bonus money can be paid to high-producing sales executives. How big is the step from that level of arrogance to the belief that they can also go one step farther in establishing wage and stock limits for all Americans?

NEW SENSITIVITIES TO THE INTANGIBLES

The tendency to ignore the intangible elements of our Capital Compound has been with us a long time. This can no longer continue, and there are evidences that our creative initiative and spiritual inspiration elements are becoming more openly recognized but not compensated.

Edward Denison, an economist with the U.S. Department of Commerce, has done a study to pinpoint which factors contributed most to economic growth during the period 1948-1973. The conclusion: that about two-thirds of the economic growth came about because of the increased size and education of the workforce plus the greater pool of knowledge available to the workers. (The Dennison study is cited in *The Service Sector of the U. S. Economy* by Eli Ginsberg and George Vojta, *Scientific America,* March 1981.)

Another interesting report is that of MIT's David Birch. His study, titled "Who Creates Jobs?" appeared in the fall 1981 issue of *The Public Interest.* Of particular importance were his findings that, of the nineteen million new jobs created in the United States during the 1970s (more than any decade before), only five percent were in manufacturing, and only eleven percent in the products sector as a whole. Conversely, that leaves almost ninety percent, or seventeen million new jobs, which were not in the products producing sector. Birch had this to say: "We are working ourselves out of the manufacturing business and into the thinking business."

However, we must also guard against being caught in the reverse trap—assuming that only the intangible elements which encompass knowledge are valid. It is the *total Capital Compound* of all three elements, labor, creative initiative and spiritual inspiration that must be kept in proper perspective. The silence of David Birch on this point is deafening. He is totally silent as to how he would suggest that this non-labor productivity be compensated. It only illustrates how brainwashed he is, along with millions of others, that they don't even see the inconsistency.

HIGH TECH/HIGH TOUCH

John Naisbitt also points out in his book, *Megatrends,* the reason why personal relationships and people skills are so important in our emerging age of information. He says:

> "High tech/high touch" is a formula I use to describe the way we have responded to technology. What happens is that whenever new technology is introduced into society, there must be a counterbalancing human response—that is, high touch—or the technology is rejected. The more high tech, the more high touch.

He further illustrates this with examples like the following:

> The introduction of the high technology of word processors into our offices has led to revival of handwritten notes and letters.

Obviously, the more we utilize computer driven tools for performing "human" functions, the more we want to reassure ourselves of our humanity. Consequently, there is a corresponding rise in the need for attention to our own balance between our self-image development and our skills at interpersonal relationships.

PROFIT SHARING

The first alternate form to consider is one several companies have already experimented with. This is "profit sharing," and it is often misunderstood to be another form of wages. There is, however, an important difference: profit sharing is only possible after the company has proven that its products or services can compete successfully in the marketplace.

Wages are a fixed cost that must be added to the selling price of the product or service. No profit can be distributed if the marketing efforts of the company have not been successful. The entire company's total family needs to accept the fact that they as a group carry the responsibility for profitability! The biggest disadvantage to this form of distribution is that it cannot properly represent the total contribution of the company family. In order for a person to receive a portion of the profit sharing for any given year, that person must have been present during that year. However, the *eternal quality* aspect of equity shares in a company allows for a person to receive payment even if they no longer work for that same company. Profit sharing is an improvement over wages, but it falls far short of owning equity in the company.

Today's American worker has come to expect instant results. However, one of the important ingredients that go along with the intangible elements of our Capital Compound is time. For example, the synergism that is created by the total group input is often delayed in its full impact. It might take three to five years for a given project to fully materialize. How then could an accounting department determine how much of the profits, if any, should be shared in a given year to compensate for things that haven't happened yet? It simply isn't possible. Furthermore, as we have said, if you're not around when the results of your contribution finally do start to pay off, you won't be in on the profit sharing.

The second most important disadvantage to the whole company family from profit sharing is the fact that, although not a fixed cost, it nonetheless creates a cash drain on the company and reduces the amount of available cash for research and development, market expansion, and other factors that are worth more to everyone than a little more cash to take home. However, in the first few years that a company family would endeavor to make the transition from a strictly wage compensation to a predominantly joint ownership position, profit sharing could serve as an intermediate step. So, at best, profit sharing should probably be limited to only a partial disbursement, and at that, only as a means to make up for needed cash flow to keep the "family" members abreast of cost of living increases, and so on.

THE IMPACT OF PROGRESSIVE SOCIALIST CONTROL

The possibility of Progressive Socialist control is staring us in the face. Should their ambitions for more control be realized, we will see them move rapidly to remove private property from the list of our inalienable rights.

In the eyes of our founding fathers, private property was held to be the absolute centerpiece for the "pursuit of happiness." Without private property rights the rest of our inalienable rights disappear in a fog of high sounding ideologies that do not deliver any substance. When one reads the constitution written for the evil Russian Empire during its reign of terror and expropriation of property and personal freedoms, you would think it very idealistic. However, you will not read anything about inalienable rights of life and the pursuit of happiness. To state these goals would have required the state to relinquish its claim to be the rightful owner of all private property. Without private property and the fruits from it, there is no foundation for idealism. Without private property rights there are no inalienable rights. The centrality of this principle in the minds of our founding fathers is seen in their provisions for *not* establishing a system of property taxes. Such taxes came much later and only then because of the incursion of many Progressive Socialists in the mix.

Under the control of the Progressive Socialists, America will cease to be America. Ultimately the change would be so encompassing as to defy description. The outcome would most likely be a rapid decline in world economic standing and then a reduction in our ability to protect ourselves on the world stage of the rising Muslim revolution.

Most likely we would become a dictatorship, not of the old European models, but more like the early models of Muslim dictatorships. Our Constitution would languish and decay in a backroom of the Library of Congress, never again to be the centerpiece of inspiration for individual freedom and private property, such as is stated in our inalienable rights of life and the pursuit of happiness.

SUMMARY

Our future possibilities under the growing success of our personal Capital Compound and that of our Joint Capital Compound would be impossible under the control of Progressive Socialists.

Just like the ancients before us, we bargain for better compensation for our work and services from those we serve. Yet, just like the ancients we have not learned who we are and how we contribute and actually control the crucible of the creation of wealth. Like the ancients, we are still bargaining—or perhaps we should say *begging*—for more cash upfront on our wages. This phenomenon now transcends millennia and yet we still persist in our belief that wages are what it is all about.

Although we might still be confused and unconvinced, technology is not. It's time to wake up and smell the flowers. Our Personal and Joint Capital Compounds is doing exactly as God created it to do. It's time to ask forgiveness for being so shortsighted and egotistical. Yes, egotistical: by insisting that it is all about wages we are actually saying that we don't believe that God has gifted us as He says He has. That's OK for those who are atheists but, despite the many faults of America, being an atheistic nation is not one of them. Certainly we do have some, but they are few in number. So what does that say about the rest of us?

Let us say this: Arise America, rebuild your God-given Capitalist foundations. Do not look back, and do not look to the right or to the left, look straight ahead. Thank God for the undeserved blessings of our intangible elements which are gifted to everyone, both believers and nonbelievers!

Chapter Eight

Arise America, Rebuild Your God-Given Capitalist Foundations

At this juncture in our presentation of this book, we are compelled to assume that our readers are much more prepared to address the challenge of rebuilding our God-given capitalist foundations. America has always led the world in the personal freedom to vote and to select ones leaders. All around the world people long to this day to be able to exercise such liberty. Unfortunately, "we the people" are responsible for allowing our beloved America to slump to the lows we see today. Likewise it is up to us to reengage and to revitalize the grassroots of millions who want and need so much to remain free and to give their children and grandchildren a future of even more personal freedom, financial security, and overall quality of life.

TEA PARTY MOVEMENT AND GRASSROOTS POLITICAL ACTIVISM

SIZING UP THE COMPETITION

Our beloved America is blessed with a large majority core of men and women of faith and hope. This core group, according to numerous polls, comprises *over sixty percent of all Americans.* These are the men and women who faithfully serve their families in the exercise of their personal labor element of their Personal Capital Compound. My studies show that no more than three percent of these same people are actually compensated for the other two intangible elements. These are men and women of courage and integrity who are not afraid to stand up for their rights and the rights of their fellow Americans. They come from many different backgrounds and ethnic groups. They are the descendants of pilgrims, pioneers, and immigrants from many different countries. They form the largest group of family leaders in America. The majority of small business owners come from this group of people. They are stockholders in small and major corporations, executives in all industries, workers in construction, manufacturing, and information technologies. Also fishermen, farmers, laborers of all levels industry wide teachers in many private and public institutions and those from the wealthiest to the most humble means come from this group. All are strong patriots. This book is written especially for this group of people!

Those individuals, who make up the Progressive Socialist movement, whether as main players or just followers, are not likely to read this book. If they do, it could convert them. If not, it will certainly alert them to the fact that, as Yamamotto said after the Pearl Harbor attack: "I fear all we have done is to awaken a sleeping giant!"

In contrast to this great blessing for our country, there are others who also participate at one level or another in all of the above positions within the American culture. These unfortunately are not a blessing. The most striking group is the Progressive Socialists who comprise *less than twenty percent of the total population* but who occupy *over ninety percent* of our academic positions, print media positions, TV and radio media positions, arts and theater positions, public school positions, civil servant positions, organized labor positions and every other position imaginable. Their positions are not by accident. They are following the core instructions of the Communist Manifesto to occupy such positions. Not because any one person or group of persons told them to, but because it is taught in all of our universities that these are the *"positions of change."* They are simply following their core ideology, Progressive Socialism.

The good news is that all one hundred percent of these Progressive Socialists comprise a much smaller percentage of our entire population than would be assumed with only a superficial examination. However, when measured by the size and amount of influence that they have wielded and continue to wield with increasing strength, one would think it was they who comprised over sixty percent of our nation's population. This means that "we the people" have the power to vote them out.

REPAIRING OUR SHIP OF STATE

America is also blessed with means and powerful tools to perform any number of corrective actions to improve her success for the Constitutional Representation of all citizens. The most powerful of these tools is the one we exercise every time we go into the solitude of a voting booth. Our voting rights are like a powerful single shot weapon. Yes, we only have one legal vote for each contest. So what is the value of voting if we only have one for each contest?

Let me answer that question with another question. How much value would you put on the shot of just one of artillery cannons mounted on the Battleship Missouri? Remember, the Missouri has sixteen-inch cannons that shoot a two-thousand pound trajectory accurately over more than twenty miles. You see, it is not how many shots our gun has, what matters is how *big* our gun is.

Do we appreciate and know just how big and powerful are each one of our votes? With our single shot voting cannon, we have the power to raise up leaders and to put down leaders. We have the power to honor good leaders with continued opportunity to serve, and to cut the legs out from under those bad leaders, those who do not hold to their commitments to serve and protect the Constitution.

Our single shot voting cannon is so powerful that even before the upcoming elections this fall there are dozens of politicians from both parties who have measured the rise of grassroots movements, such as the Tea Party Movement. They have concluded that it would be best for them not to run this fall. Democrats and Republicans alike are pulling out of the competition for office. Those of us who know what's happening in our government today and those who trust us that we know have not even fired the first shot. Yet, our *enemies* are pulling out of the fight.

Let me be clear about this. In most political contests I would not characterize those of opposite opinions to be my enemies. However, when we are talking about Progressive Socialists, you are not just talking about a politician with a different opinion. You are talking about self-declared enemies of everything our Constitution represents and the freedoms it guarantees us to have if we will follow it and defend it. Make no mistake about it: Progressive Socialists, despite their party affiliation, education, job description, birth, and ethnic origin, the size of their bank account or any other means of identification are, by their own self-declared definitions, all America's enemies!

This latest round of arrogance was demonstrated by forcing the most horrible legislation ever, the Health Care Bill, on the American people who had already shown by a wide majority that they did not want it. This proves that they are our enemies. As we write this book, hundreds of legal briefs are being prepared by America's finest Constitutional lawyers, soon to be filed against this horrendous, unconstitutional legislation attempt. Any enemy of my beloved America's Constitution is my enemy. How say you?

LOADING OUR SINGLE SHOT VOTER'S CANNON

Loading our single shot voting cannon is synonymous with knowing why we are voting for a given candidate, regardless of their party affiliation. Party affiliation is the least of our considerations. It only takes a cursory review of past political candidates from both parties to realize how misleading party affiliation can be. Personally, I am most likely to vote Republican, but I am always on the lookout for the right man or woman, regardless of party affiliation. Our most recent election cycles have brought special attention to the fact that many who claimed to be conservatives were everything but. Then there have been those from the self-declared Progressive Socialist party, the Democrats, that were much more conservative than their Republican counterpart.

Understanding the vocabulary of politics is one of our strongest tools for properly loading our cannon. For example, how many of us understand the term "moderate"? When I hear the term moderate I always want to ask the question: "Moderate compared to what?" A .22 caliber bullet in the shoulder is moderate compared to a .44 Magnum slug: the .22 only makes a small hole, whereas the magnum would blow my shoulder completely off.

The truth is, I don't want either a hole in my shoulder or to lose the shoulder completely. So, if I'm looking at what appears to be an option for a less evil politician—such as a .22 caliber bullet—and he/she calls themselves "moderate" my antennas are immediately up and twitching!

In the interest of transparency I will include names from time to time: John McCain, in this instance. John is a great American hero and has suffered things that none of us would want to suffer. Without taking anything away from his heroism there are other parts of his life that do not measure up to the same sacrificial reality that he demonstrated in the Hanoi Hilton.

For example: among many other illustrations that could be presented where he has shown his true colors of Progressive Socialism, the one that stands out for me was his postponement of his presidential campaign to go back to Washington and make a great fight for real conservativism—right?

Wrong. He went back to Washington and folded up like a cheap suit, joining his fellow Progressive Socialists in passing the bank bailout bill. When the chips were down he showed his true colors. This is always the case with "political moderates." I deliberately differentiate between political moderates and just being moderate about something. The term moderate within the political context is used specifically to give the impression of someone who is less extreme in their views, therefore eliciting sympathy and support. The truth, in comparison, is actually quite shocking!

We propose that in matters of principle and character there is no such thing as moderate! Please allow me to put it in an easier context to grasp, so as to better understand to what I am referring. If I say to you that I am a man of moderate integrity, moderate character and moderate honesty and I would like for you to go into partnership with me in a very big deal involving a lot of money and responsibility, how do you feel? Have I inspired you to respond positively, or are you having second thoughts? If you aren't having second thoughts, I would like to tell you about some Arizona beachfront property that I happen to own.

Think what might have happened if John McCain were not a Progressive Socialist and would have gone back to Washington and stood up to Bush and all of the other Progressive Socialists—those economically illiterate useful idiots—and had refused to join forces with them. In hindsight it is easy to imagine that he would have cut a new place for himself in history, possibly even a seat in the White House. Now, with his Progressive Socialism in full view, it will take a miracle for him to regain even his Senatorial seat from Arizona, as J. D. Hayworth is poised to give him the run of his life.

Consider this: when J. D. Hayworth first announced his candidacy in January 2010, there were twenty-six points separating him from John McCain, according to the Rasmussen polls. Now, after less than four months, and despite two million dollars in negative ads by McCain, they are separated by only five points.

AIMING OUR SINGLE SHOT VOTER'S CANNON

Loading our voting cannon is only the beginning. The most difficult task is to aim our cannon. By aiming, we are referring to taking self-conscious control of our actions as the date for voting comes closer. Our single shot voting cannon is only representative of our personal Capital Compound: what we need to take down the enemy is a full blown fuselage. Even though we have possession of only one shot for our single barreled cannon, we work and operate within a team of many other owners of a single shot voting cannon, i.e. the millions of grassroots Tea Party members and sympathizers.

When we prime not only our own cannon but also assist in the priming of other cannons, we then have the power of a tremendous full blown fuselage. The battleship Missouri has nine sixteen-inch cannons, whereas we have countless thousands of cannons with much larger bore than just sixteen inches. Wow, now we are talking about real power and real shock and awe. It now becomes obvious to us that we do not want to work or to think as just one vote or as just one person: we must learn to think and act as a huge team, moving ahead in lockstep, focused on the same goals, taking back America and our Constitutional rights!

As a team we realize the power and penetrating force of our Joint Capital Compound elements. We now have greater initiative and creative inspiration and as a result, we have much greater synergism. This is equivalent to magnum force in ammunition terminology! Our personal Capital Compound represents one vote, but our Joint Capital Compound represents all of the votes that we can convince to put our same undesirable coordinates into a mass aiming process. Together we are like a smart weapon that cannot miss!

OVERCOMING BUCK FEVER

Buck fever is that moment of indecision when a hunter has his prey in the crosshairs and then freezes for no apparent reason. Under the influence of buck fever, the hunter will either pull the trigger and miss or hold the shot for so long that the target moves off and the shot is spoiled. It is agreed among the experts that buck fever is the direct result of either a lack of preparedness of a lack of self-confidence.

Preparedness is the best solution to a lack of self-confidence. Since we all need more self-confidence, it also says that we all need to be more prepared. For example: If I am asked to give a one hour speech, I can expect to be ready within only a short time to prepare. However, if I am asked to give a five minute speech, I know that I will need a considerable amount of time to prepare. The challenge is when we are asked to say something or do something important and to do it in a short time.

MOTIVATING OUR JOINT CAPITAL COMPOUND

We build confidence by keeping our tasks within our capabilities and by organizing them so that they are easy to recall and to implement. One of my many blessings is that I have had the privilege of working extensively in finance, sales, marketing, business management, capital acquisition, and political science.

As a serious student in the study and execution of these disciplines I have taken on the responsibility, first and foremost, to write this book especially for your benefit. To further this pursuit for your benefit, I have compiled a list of bullet points that I encourage you to copy and to put in your briefcase, car, RV, private plane or any other means of conveyance you have access to. Pull them out often, read them, and commit them to memory. You will find that in a very short time you will be using these bullets in your daily conversation with others. This introduction and exchange will open the door for you to share your thoughts on our beloved country's greatest needs and at the same time to encourage others to do as you have done: to learn and to get involved.

Socialism destroys the very essence of our motivation.

1. Hope is essential. Without the hope of entrepreneurialism, private property, and profits, there is no wind to fill our sails. We are left adrift in a sea of Socialist sharks circling our fragile craft in hopes of one last meal from our starved, emaciated bodies.

2. Government that is Socialist in its worldview is a cancer to all entrepreneurial enterprise.

   a. A small business owner takes one dollar and invests the whole dollar in the business.

   b. All business expenses are void of taxes; they are not part of the taxable gross profit.

c. Contrary to the small business man, the Socialist-minded government takes one dollar in taxes or new debt, and spends eighty-three percent of it on administration, leaving only seventeen percent to "invest" in the objective of their Socialist intent. In our current situation the government is telling us that they are going to invest billions of dollars into small business loans. To do this they need six billion dollars in taxes or new debt for each one billion dollars that they loan to small businesses. Remember only seventeen percent of each tax dollar will ever reach the intended purpose.

4. Socialism is a device of unholy greed for power and control to incite jealousy and hatred against all those accused of being the evil rich. Most of the individuals now making $250,000 or more annually are small business owners, therefore they are by definition the evil rich.

a. Socialism assumes as absolute the injustice of profits, therefore all entrepreneurs—read small business owners—are guilty of unjust profits.

b. Socialism and Communism, in all of its many forms, are one hundred percent compatible with the worldview and thinking of all union bosses.

c. Remember the phrase: "All workers unite." This has always been the clarion call of the Communist and Socialist parties. It is no accident that our Socialist-minded government is in bed with organized labor. They both view profits as evil.

d. Marx considered "profits" to be an expression of labor's fruit and that it belonged to labor and not to management, owners, or investors.

e. Organized labor as represented by unions is inherently self-destructive and anti-labor. This phenomenon occurs every time a union is successful in gaining increases in their member's wages and benefits. These increases become a new factor in the expense column of the business affected. The business owners then use this increase in overhead to spur their decisions to increase the implementation of more advanced efficiencies in machinery, infrastructure, automation, and organization. These management decisions automatically lead to layoffs of—believe it or not—union members. This is one of the negotiation methods used by both the union and the business to pay for the increases in wages and benefits. It is customarily accepted by both that a certain number of the lower seniority union members will be laid off. Over time this has reduced the percentage of union laborers in the USA. It is important to note that in the case of our Socialist-hinking government, the civil service sector is protected against this aspect of their contract increases. The government civil service sector does not produce a marketable product that can be significantly economized by improvements in automation and technology. Even when such improvements are made, it never leads to a reduction of overall government employees or cost factors.

f. It is no wonder that most of the unionized workers of today are government employees. Local, state, and federal including all of the teacher members of the NEA, and other teacher unions on the national, state, and local levels are government employees.

g. In the case of our private enterprise, just imagine the scene in the backroom of such negotiations. The labor representatives who claim to be "pro-labor" are willing to throw a certain portion of their members under the bus in order to enrich the majority. This is what happens in every contract increase. There are always a certain number of members that they are willing to sacrifice. You never get to hear from these members. They are incognito. I wonder how many of them are happy with the way they were represented. Was that the change they wanted?

5. The Socialist-minded government addresses its message of unholy greed and power grabbing to a postmodern society that is primarily divorced from the reality of entrepreneurialism, private capital, and profits.

   a. The wage earner is told that the rich have more than they deserve and that if the workers will vote them into power they will take that undeserved profit and redistribute it to those who deserve it more.

   b. At the birth of our nation ninety-five percent of the population was engaged directly in entrepreneurialism, private capital, and profits. For this reason they only allowed property owners the right to vote. In a culture and world dominated primarily by entrepreneurs who were property owners, who understood the absolute necessity of private capital and profits, it was just and proper to give the right to vote to them only.

   c. In our current postmodern world we consider everyone, both the financially informed and the financially ignorant, worthy of the right to vote. In and of itself, this is not a bad thing. Consider the fact, however that it is estimated by all schools of financial experts that only ten percent of Americans understand entrepreneurial finance, private capital, and profits. Personally, I believe that there are less than five precent who understand these financial realities and that those who are ignorant of such knowledge and experience numbers over ninety-five percent. When this is compared to the reality of our early founding fathers and their culture of ninety-five percent who were directly involved in entrepreneurial finance, private capital, and profits, we can understand what Thomas Jefferson meant when he warned us of how the nation could go bankrupt.

d. Thomas Jefferson warned that when the general populace became aware of their opportunity to vote themselves payments from the national treasury, the nation would go bankrupt. Our current Socialist-thinking government is now determined to take over the major centers of business of the economy and to also control the labor unions.

e. The labor union bosses of today are no more than extensions of the power hungry politicians and political centers of self interest in high places.

f. Barak Obama has made it crystal clear that he despises profits in the private sector and that he will stop at nothing to gain control over them for his own Socialistic dreams and purposes.

g. However, it is impossible for him to accomplish this without the support of willing accomplices, i.e. union bosses, Socialist thinking politicians, and ignorant and financially illiterate voters who are convinced that they can steal from the profits of private entrepreneurs without destroying both the economy and themselves.

6. Socialism is a moral and mental disorder.

a. Socialism is a moral disorder because it accepts a concept of stealing as honorable.

b. Socialism is a mental disorder because it produces criminal acts which destroy both the victims and the perpetrators.

c. Only a diseased mind can embrace a concept that blatantly disregards the truth. Disagreements over what is truth are normal. On the other hand, a mind that chooses rather to disregard all morality and to self-consciously engage in acts that commit both suicide and murder at the same time is deranged.

SUMMARY

PROPOSALS AND SOLUTIONS

In the following Chapters Nine, Ten, Eleven, Twelve and Thirteen we take on the more difficult task of actually presenting proposals and solutions. Based on all that has been presented up till now, I feel confident that all of our readers will find the next four chapters very exciting and disturbing at the same time. It's exciting, because you will discover new wealth and financial security for you and your family. It's disturbing, because you will also discover the inevitable difficulties that are awaiting us just over the next hill.

My fellow Americans, what more can we say at this point? You know the issue better now than anyone else who has not read this book to this point. Now that you know, our challenge is to actualize ourselves to use all means of honorable and honest principles, to begin the process of rebuilding our God-given capitalist foundations. You can now appreciate as never before just how powerful is our one vote. We have the power and we have the creative initiative. We must act at the local level, the state level, and the national level. We have the power of the vote, let us use it!

Chapter Nine

Beyond the Fear of Failure

In his first inaugural address in 1933, FDR said "We have nothing to fear but fear itself." This statement was in reference to the depression the country was suffering, much like the country suffers today. Americans are a people of faith, not a people of fear! By God's grace and providence we have overcome every enemy that has come against us since the days of our earliest colonies. We have been blessed with a spirit of overcoming, a spirit of valor and courage, this is our legacy and we will rise up and overcome our enemies.

HOPE OVERCOMES FEAR

Faith is the substance of things hoped for and the evidence of things not seen. - Hebrews 11:1

Our country's most respected pollsters have recorded that over eighty-three percent of Americans believe in God. There are no polls telling us how many believers there are in the Tea Party Movement, but having been personally inspired by their character and integrity, I believe that more than eighty-three percent of them are men and women of faith. Faith and hope are key parts to our Personal Capital Compound. Without faith and hope, the intangible elements of our Capital Compound are empty and void. In contrast, Progressive Socialism is a cancer to all entrepreneurial enterprise and sucks the life out of our faith and hope.

The character and faith of the Tea Party Movement should come as no surprise. It is this same character and faith which prevails in the larger majority of all Americans. Without a doubt, our nation's most valuable national treasure is our huge number of fundamentally Christian people. Our nation was founded on Christian faith and principles, so much so that every state constitution pledged allegiance and commitment to God when first drafted. Every state constitution clearly sets forth language that appeals to God for strength, wisdom, and His providence to succeed.

Unfortunately, this same national treasure has allowed itself to become delusional and misguided, with respect to the personal political responsibility of the people that make it up. For all practical purposes they no longer participate in the political process. Consequently, we now find ourselves in the political war of our life and that of our beloved country. Let us pray that this new grassroots revolution will awaken these many millions of politically inactive Americans.

Without the confidence of our faith and hope we lose the focus for our vision of success for our children, our grandchildren, neighbors, and the country as a whole. Socialism destroys the very essence of our motivation. Hope is essential: without the hope of entrepreneurialism, private property and profits, there is no wind to fill our sails. We are left adrift in a sea of Socialist sharks circling our fragile craft, in hopes of one last meal from our starved, emaciated bodies.

It is critical now, at this time of our country's greatest political crisis in history, that all men and women of faith step up and take serious aim with their single shot voter cannon, with nerves of steel and open fire. Without actually making the vote (pulling the trigger) our voter cannon has no power to overcome. Our first commitment and duty to ourselves, our families, and our country is to be totally committed. Come what may on voting day, we will do whatever it takes to get to that voting place and cast our vote. Everything we are learning and everything we are saying or doing today that falls short of making that vote becomes meaningless!

I look forward with faith and hope to this coming fall, when I will see the greatest voter turnout in the history of America. Not just in my town or in my state, but all across America. There has never been a midterm election that has held so much importance for so many people as the one that we will all celebrate this coming November. I believe that this coming together of our national Joint Capital Compound, as expressed in our upcoming voter turnout with our massive fuselage of firepower, will cause a change so big and so powerful that it will go down in history as the greatest paradigm shift in political power to date. This will resemble the severest of earthquakes, of such political seismic shock that it will ripple from west to east and from north to south. It will be a shift from blatant Progressive Socialism to true conservatism so strong that it will rank alongside the signing of the Declaration of Independence!

Believe me, we will see this happen and we will all be amazed, even though we will not be surprised that it happens. The strength and complete national impact will amaze not only us, but the whole world. What Japan did at Pearl Harbor to awaken the "sleeping dragon" as Admiral Yamamotto said, will seem to have been no more than a child's firecracker going off, compared to the explosion coming this November.

SUMMARY

Having said that, let me hasten to add that I am equally convinced that we must increase and greatly deepen our understanding concerning the underlying ideologies of the enemies of America and the issue of our contention for wealth and its equitable distribution! We must deal not only with the *first step*—which is to create the opportunity for this great political shift in power—but with the *second step*, the issue of wealth and its equitable distribution. Then we will witness a new era for a greater America that will surpass anything in our history to date.

If we fail to do this, however, we will be no different than the conservative voters of the generations that preceded us. Time after time they would see the enemies of America destroying her Constitution and our freedoms, they would rally, they would win (at times even with great majorities) and then they would go back to sleep. Fellow Americans, we who are conservatives and who cherish our beloved country, our Constitution and our freedoms, no longer have the luxury to go back to sleep. Our personal freedom, the freedom of our children and their children's children, hangs in the balance! Freedom is a 24/7/365 struggle. We do not want to repeat the same mistakes of our predecessors.

We must allow history to teach us and let our current struggle be the beginning of our fight, not the end. We have much to accomplish to overcome our ignorance on the issues of wealth and its equitable distribution. Our challenge is to do this and to remain faithful to the character and integrity set forth in our Constitution.

If we are committed to both the first step and the second step, the God of our faith and hope will strengthen us and enable us to do mighty deeds and to overcome mountains of criticism, of abuse, of characters maligned and yes, of fear itself! For all who are committed to this two step plan of attack, you are welcome to read all of the following pages, humbly offered for your benefit to deepen your understanding of the issues of wealth and its equitable distribution.

My fellow Americans, we are challenged as never before, even more so than in the depression of 1929 or the Second World War. We face a relentless enemy, but our enemy does not know who we are. In their minds, we are weak, despicable, lowlife trash and without enough brains to come in out of the rain. The arrogance and ignorance of our current Progressive Socialist leaders is both historical and repugnant to every decent American citizen. Despite their disdain for us, we are more alert, more informed, and more prepared to take this fight to the ballot box than they can imagine!

Arise America rebuild your God-given capitalist foundations!

Chapter Ten

The Issues of Wealth and
Its Equitable Distribution

Marilyn and I are parents of six and grandparents of twenty-four. We are always concerned about the future of America and the protection of our immediate family and for many generations to come. We are not unique: each of you, to one degree or another, have the very same concerns for your own families and their future generations. Our concerns, yours and ours together, illustrate that the risk of what does or doesn't happen in the future of our beloved America is of great importance to us all! Protection against risk is a natural response to one's concerns of danger. However our greatest risk is actually expressed in who holds the power to change our lives and to threaten our family's freedoms to actively progress financially, morally, spiritually, and to be free of terrorism, both from within and without.

SECURITY

It is widely reported that a middle-aged lady once said to President Eisenhower, as he was campaigning in the Midwest:

> "Mr. President, I want security, I want to be so secure that I can't move!"

The President responded, "No you don't ma'am; only people in solitary confinement have that kind of security!"

A BASIC DESIRE

Security is one of our basic human desires. It is grouped with desires for such essentials as food, breath, sex, protection from the weather, and companionship. With all of our basic desires, we have a capacity for extremism, and security is no exception. As we consider our history of attitudes toward security and risk we need to be sensitive to our extremist tendencies wherever they exist. Risk management is a virtue that serves a great purpose in the creation and distribution of wealth.

It is vital in our treatment of such important matters as wealth and it's distribution to look candidly at our attitudes about risk. The creation of new wealth has few, if any, guarantees. Yet we are not allowed the option of existing without some relationship with the phenomenon of wealth and its distribution. On the one hand, we can certainly be forgiven for not wanting to expose ourselves to needless and foolish risk; on the other hand, we do not want to be as secure as the person reduced to solitary confinement.

SECURITY WITH FREEDOM

The ideal situation is that everyone has true, practical, and affordable security relative to subsistence while at the same time enjoying complete freedom in the pursuit of political, religious and personal goals.

The Communist philosophy says that, in order to merit security from the state, the individual must give up his right to as many personal freedoms as the state should deem necessary in order to maintain the security. On the other hand, our republic with inalienable rights allows us the freedom to experience risk.

Our security in matters of politics, personal freedom to pursue happiness, the right to own property and to be protected from the state is made possible under the laws that give us the right to correct our mistakes or to endorse our victories. If we should agree as a voting society of people, that a certain attitude, restriction, or opportunity be expressed in the form of a new law, and that law as expressed should prove to be unacceptable by popular demand, we can change it. Or, if such a law or laws should prove to be very practical and to serve our needs well within the structure of the Constitution, then we can choose to support and retain them. Thus, we have a measure of control over our risk.

This is sometimes a frustrating process, but history continues to prove that it is the most effective method of providing the maximum amount of political and personal freedom for all. Our only real danger is the lack of vigilance. When we lose our vigilance we soon become delusional, giving us a sense of false security. We must learn a great lesson from this and allow faith and hope to prevail in an active involved way, and not in a passive and uninvolved way!

This action keeps us close to our everyday realities, gives us a clearer perspective of acceptable risk, and protects us from becoming delusional about our security and falling asleep as our predecessors have consistently done.

SHARING RISK

In the democratic process as embraced within our republic, we see that risk in political and personal freedoms is divided up among all of the voters. Our freedom to speak out and to assemble for discussion of ideas and proposals is synonymous with our security and our protection from risk.

If we can understand the blessings of the democratic process and the security we enjoy through our republic with a representative form of government, we then can surely understand the value of every person owning a portion of the instruments of production of our Joint Capital Compound. It is through such a broad base of ownership protected within the liberties of our republic that we can address the problem of financial risk with optimism. There is no other country in the world formed under such a powerful constitution, with liberty and justice for all. However, before we can fully appreciate the various ways that our Joint Capital Compound ownership will afford us, we need to review some common attitudes to avoid risk and present trends in our response to it.

AN AGELESS BATTLE

Securing against future risk is a principle that dates back many thousands of years. Drawing from Genesis 41:1-57: It is recorded that Joseph (the son of Jacob, a Hebrew patriarch) was appointed by the Pharaoh of Egypt to oversee the storing of grain during seven years of good harvest to prepare for the shortages of seven years of famine. As early as 1,700 B.C. the Babylonians had a form of credit insurance called The Code of Hammurabi. Under one of the provisions of the code, if personal misfortune made it impossible for the borrower to repay his debt, he was relieved of his obligation. The borrower paid extra for this protection.

The concept of insurance to help protect against risk has been used by different societies ever since. In Roman times there was an organization called *Collegia Tenuiorum*. This organization provided burial insurance for its members. Surprisingly enough, the membership consisted of slaves and wage earners. Another organization, called simply the *Collegia*, provided the same service for military personnel.

During the Middle Ages, craftsmen formed organizations called guilds, the forerunners of labor unions. The guilds offered a similar type of burial insurance for their members. In addition they also offered fire and theft protection.

The first fire insurance company in America, called *The Philadelphia Contributorship for the Insurance of Houses from Loss by Fire*, was founded in 1752. Benjamin Franklin was instrumental in helping to found the company, and also helped to found the first life insurance company in America, *The Presbyterian Minister's Fund,* in 1759.

People of property have always sought ways to protect it, from digging moats around their castles to insuring goods carried across dangerous oceans. However, like the slaves, wage earners, and soldiers of Roman times, the common people have rarely had more to insure than their lives or a few possessions of minimal value. Not until after the turn of the century did many Western countries develop the diversified insurance programs that we have today, largely in response to the rising incomes of workers. These programs are primarily designed for the needs that are associated with a laboristic form of distributing wealth through wages and other forms of cash payment. The creative insurance programs possible in a society that practices an equitable ownership distribution of our Joint Capital Compound profitability have not yet to been developed.

We can clearly see from history that, when given the opportunity, mankind has attempted to protect himself from risk. Even though we consider this to be normal, the basic concepts of laborism and the laboristic distribution of wealth through wages and benefits have created many extremes.

NO SECURITY IN LABORISM

One of the underlying seeds of decay in the laborist concept of distribution is the idea of security through labor, or "job security." We have unwittingly adopted this idea from Marx and Engels' Labor Theory of Value. They could not imagine fruitful and just distribution of wealth, not even through wages as adopted by the Free Market societies. In their view, the solution was to expropriate all wealth and all private property and to turn it over to the state. Their solution was one of *ownership* alright, but not for individuals. In their minds the only owner was to be the state. Thus, the workers would be guaranteed an income for labor, and the workers' escape from risk would be complete. Well, we all know what disastrous results that theory brought about. We have a world full of examples of such would-be ownership!

As we pointed out in Chapter Three, Marx and Engels probably didn't believe that their radical solution would destroy the individual's opportunity and freedom of political action. Fortunately for us, we have almost a hundred years of Communist track record by which to measure the results. In contrast, the workers of democratic countries, and even those of the free market dictatorships of the world, try in vain to acquire this guarantee of security and freedom from risk through "labor contracts." This has resulted in the development of our current primary distribution through laborism, and the unfortunate expropriation of the intangible elements of our personal Capital Compound, creative initiative and spiritual inspiration.

Our society's adoption and development of insurance programs such as Social Security, Unemployment Compensation, Worker's Disability Compensation, major medical coverage, and retirement plans are further evidences of our attempt at securing ourselves from financial risk through laborism. As they are now administered, all of these social and financial insurance programs are dependent on the idea of Full Employment for their success. All of these attempts at personal security have all had their beginnings from Progressive Socialists of one description or another.

Beginning in 2008 we saw the loss of equity and value stored in 401K programs and other financial vehicles designed for retirement. All of these plans, even though there was significant use of stocks and securities from around the world, lost value rapidly. The underlying dependency for success in all of the stocks and securities was and continues to be based on consumerism, which translates into steady employment, which in turn gives "value" to the 401K. When the chain of events is reversed the outcome is predictable.

REALITY DEMANDS A NEW METHOD

In the not too distant past, we suffered dramatic economic reversals in our "smokestack" industries, when literally hundreds of thousands of workers lost their financial security based on laborism. The only thing that the workers could trust to stay on record and be picked up again was their contribution to Social Security. And, with the government's mismanagement of that fund, who can get excited? These dispossessed workers lost their medical benefits, their retirement benefits and, for many, their unemployment benefits ran out before they found a new job. As a result of these and our most recent developments, homes have been forfeited. Automobiles, recreational vehicles, and televisions have repossessed and every other conceivable consumer item has been lost. All of these losses are directly attributable to our extreme attitude toward demanding wages—cash payment—upfront.

For those of us who might not have suffered any such reversals as yet, we could have a serious lack of sensitivity to the danger we face. The signs are evident and plain to see when examined in the light of the Capital Compound Theory of Value. Unless American workers accept the challenge of arming themselves with information and working in unison to reverse our present direction through the legislative process, literally tens of millions will suffer such personal losses! The latest figures on unemployment show over fifteen million people are out of work and the list keeps growing.

EVERYONE IS AT RISK

The fortunate few who presently own equity in the instruments of production should not believe for one moment that they will be spared. Should our present state of financial and economic illiteracy continue, and thus propel us forward in our present direction toward financial self-destruction, then they too will ultimately suffer great personal losses. When tens of millions lose their property and their personal possessions because of laborism's inability to keep up with our developments in high technology, can we not imagine the public resorting to voting in massive controls, and/or centralization of ownership of private property to the government? Let's be honest with ourselves. If we are backed into a corner with no apparent hope of escape, save for massive government intervention, will we not join those millions who have also cast their vote for total government takeover?

Our founding fathers had the great advantage of their personal experience with creating wealth and its equitable distribution. They were both witnesses and participants in what had been 150 years of the most personal wealth creating period in history. From the first colonist to the signing of the Declaration of Independence, the personal wealth of the average colonist far exceeded that of the citizens of England and Europe. Our Constitution is the culmination, in historic legal format, to formalize and give to us the freedom to exercise our inalienable rights.

But even so, all was not perfect. In fact, by the signing of the Declaration of Independence, there was strong evidence that the seeds of laborism had taken root in the heart of colonial America.

We see how far we have strayed from those days of a more complete expression of our Capital Compound and a more equitable distribution of the wealth created. May God grant us the grace to raise such a banner of concern and opposition to our present state of affairs as to drive out the possibility that such a choice would ever be presented. For, should such a tragedy actually materialize, we would find that there would be no lack of evil politicians who would be willing to step forward and propose such sweeping changes in our republic and our freedoms.

COMPENSATION WITHOUT RISK

By far the majority of Americans believe it is their right to demand a form of compensation and distribution that is without risk. It is obvious that from the very beginning the workers did not develop, nor were they encouraged to develop, a sense of confidence in the production and free market process. Their individual and collective hue and cry has been for cash payment upfront, with few exceptions. Certainly stock options, and stock as a form of payment and distribution, have found little place among the masses of workers. Such creative methods of distribution of wealth have most generally been limited to executives and middle management.

Indeed, the workers of America have "bargained for a penny," and have left the greater portion of their Capital Compound production of wealth intact in the hands of their employers. Imagine if you can what an improved position the workers of America and of the world would have today if they had only known enough to bargain for a portion of the *total* productivity of the Joint Capital Compound. The benefit of generations of technological developments would be represented in literally tens of millions of families today. The passing of such wealth from generation to generation would have revolutionized the broad development in private ownership, hundreds of times more than we have experienced under laborism.

"LABORISM" HAS INCREASED OUR RISKS

As Job said, "What I have feared most has befallen me." The lack of an understanding of the dynamics of the Capital Compound and its synergistic component, coupled with a fear of the production and free market process, means we now face more risks than at any time since the beginning of the Industrial Revolution.

What we thought would provide us with a riskless society, has resulted in the highest form of financial risk—the risk of total loss. I'm reminded of the days of President Johnson, and the Great Society's "War on Poverty." The lion's roar has turned into a puppy's whimper!

What we actually got was acceleration in the wrong direction that has not been corrected sense that time. We resemble a financial avalanche careening down a mountain, picking up width, depth, and speed as we go.

We are now in the grips of this avalanche of government, a government fat with greed, arrogant, power-centered, and financially, morally, and directionally out of control. We must admit we are bankrupt. We must admit our mistakes and deal with it now!

CPAC-2010 KEYNOTE SPEAKER

Glenn Beck, a noted author and host of the Glenn Beck Show on Fox News gave a resounding keynote speech filled with many poignant statements worth remembering. Here are a few of those statements:

1. Jobs are not in the Bill of Rights.

2. All movements have a similar problem—hijacking from the five percent who are extremists on the fringe.

3. The best way to transport democracy is by example.

4. Our future is not cast in stone.

5. Cut spending on entitlement benefits.

6. Cut taxes, cut earmarks.

7. Medicare is a slave to inflation and it should be a slave to life expectancy.

8. We are not past the worst. We are in for an economic holocaust that I've been warning about for four years.

9. Major tax rates are scheduled for Jan. 1, 2011.

10. The average annual pay in government is 72,000 and in private sector 40,000.

CONSTITUTIONAL CONSERVATISM: A STATEMENT FOR THE TWENTY-FIRST CENTURY

Following is the Mount Vernon Statement, a statement posted online by Family Research Council, and being signed by millions across the country.

> We recommit ourselves to the ideas of the American Founding. Through the Constitution, the Founders created an enduring framework of limited government based on the rule of law. They sought to secure national independence, provide for economic opportunity, establish true religious liberty and maintain a flourishing society of republican self-government.
>
> These principles define us as a country and inspire us as a people. They are responsible for a prosperous, just nation unlike any other in the world. They are our highest achievements, serving not only as powerful beacons to all

who strive for freedom and seek self-government, but as warnings to tyrants and despots everywhere.

Each one of these founding ideas is presently under sustained attack. In recent decades, America's principles have been undermined and redefined in our culture, our universities and our politics. The self-evident truths of 1776 have been supplanted by the notion that no such truths exist. The federal government today ignores the limits of the Constitution, which is increasingly dismissed as obsolete and irrelevant.

Some insist that America must change, cast off the old and put on the new. But where would this lead -- forward or backward, up or down? Isn't this idea of change an empty promise or even a dangerous deception?

The change we urgently need, a change consistent with the American ideal, is not movement away from but toward our founding principles. At this important time, we need a restatement of Constitutional conservatism grounded in the priceless principle of ordered liberty articulated in the Declaration of Independence and the Constitution.

The conservatism of the Declaration asserts self-evident truths based on the laws of nature and nature's God. It defends life, liberty and the pursuit of happiness. It traces authority to the consent of the governed. It recognizes man's self-interest but also his capacity for virtue.

The conservatism of the Constitution limits government's powers but ensures that government performs its proper job effectively. It refines popular will through the filter of representation. It provides checks and balances through the several branches of government and a federal republic.

A Constitutional conservatism unites all conservatives through the natural fusion provided by American principles. It reminds economic conservatives that morality is essential to limited government, social conservatives that unlimited government is a threat to moral self-government, and national security conservatives that energetic but responsible government is the key to America's safety and leadership role in the world.

A Constitutional conservatism based on first principles provides the framework for a consistent and meaningful policy agenda.

- It applies the principle of limited government based on the rule of law to every proposal.

- It honors the central place of individual liberty in American politics and life.

- It encourages free enterprise, the individual entrepreneur, and economic reforms grounded in market solutions.

- It supports America's national interest in advancing freedom and opposing tyranny in the world and prudently considers what we can and should do to that end.

- It informs conservatism's firm defense of family, neighborhood, community, and faith.

If we are to succeed in the critical political and policy battles ahead, we must be certain of our purpose. We must begin by retaking and resolutely defending the high ground of America's founding principles.

OWNERSHIP HAS LESS RISK

We need to re-evaluate what we have traditionally called "risk" and focus on promoting the just distribution of the wealth produced by the total Capital Compound, and not just its labor element. Basic to this is the premise that ownership of equities is a far less risky position than is our dependence upon laborism. We must accept the fact that although there is some short-term risk in equity ownership, there are no long-term answers without it! Our colonial forefathers considered a wage earner to be no more than a chattel servant, someone who paid for their opportunity to learn a trade by giving their labor in return. The common laborers in the colonies were primarily those who had bargained for passage from the old country in exchange for their services. Their hope was to gain the opportunity to learn a trade and to gain a foothold in the new world. The remainder of the laborers of that time was those unfortunates who had lost parents and their original stake in the new world.

As the colonies became completely settled, and with agriculture being the primary form of new wealth, there was increasing pressure to open up new territories toward the west. Agriculture requires land and the growing number of colonists reached out to claim new lands for new profits.

Agriculture has always been a primary source of new wealth. Its one major disadvantage is that it requires a lot of room: it is not a concentrated form of wealth creation. Our current rapid increase in knowledge, with its technological applications, differs in that it provides almost numberless means for creating new wealth. These new sources of wealth, unlike agriculture, do provide endless expansion without the need for more space. In fact, we now experience the increased miniaturization of these technologies, further decreasing the need for more space. Technology is now reducing the amount of space it takes for agriculture, with applications like hydroponic agriculture. This, in addition to the success of technology to use new ways of processing agriculture products, is only the beginning!

The coming avalanche of super-high technology, with artificial intelligence as the centerpiece, will make obsolete all of our past thinking relative to risk as applied to job security and finances. Those who catch the vision and begin to prepare themselves for participation in collective bargaining, for the legislative process, and for creative confrontation will overcome the risks.

There is no real alternative for success. Failure could come in many forms. What we have now is failure and it will never provide us with the discoveries needed to provide us with the economic tools for a new and more successful future. Truly we must accept the risk of equity with all of its new ways of distribution.

A CURABLE CANCER

Our misguided and archaic adherence to security through laborism is a seed of cancer to more than just our financial security. Laborism is eating away at the very roots of our society: all of our basic freedoms are in jeopardy. We have clearly seen in the Communist model of centralized power the total state ownership of the Joint Capital Compound and the high price that is inevitably paid. Such is the total loss of freedom as we know it. At stake is our right to own private property, our total loss of political freedom, and the reduction of our personal and religious freedoms to a mere pretense.

Against the possibility of this tragic outcome, *security through equity ownership* is fundamentally sound, and will, given time, even cure our laboristic cancer. Once we accept the reality of our Capital Compound and insist on a distribution process that includes equity shares as well as wages, then our "cancer" will wither and die, and our freedoms will remain intact.

Although a truly riskless society is impossible, the level of security and freedom from risk that is possible through our shared ownership of all the instruments of production is much greater than what we have under laborism. As hundreds of thousands of our American workers have experienced, when your industry or job skill is displaced by high technology, world competition, or a combination of both, your financial security has a lifespan of about six months to a year. In that short period of time, you stand to lose everything you have worked a lifetime to put together. If someone could guarantee us that we will not be displaced by super-high technology, or worldwide competition that can undercut us in labor costs, then we would have few anxieties. We could go right on trusting in our present system and paying for our security and protection from risk with funds that are dependent upon Full Employment.

THE BEST SECURITY

Have you ever wondered how the owners of large equity positions in our nation's companies manage to survive massive layoffs and the ravages of financial recessions and depressions? They don't work for a wage, and they don't need unemployment insurance because they don't have a "job" in the ordinary sense. When they or any member of their family gets sick, they do not have to wonder if their "medical plan" is still in effect, or if they have turned in enough work hours to qualify for their health benefits, like so many workers have to do. If a certain branch of their company begins to show signs of weakness and an inability to keep up with technology or world competition, do they just go down in flames?

Of course not, they simply shift their capital to a more competitive position, even if it means the ultimate closing down of that certain branch at a loss of thousands of jobs. How do they manage? The answer is simple: they manage very well!

FINANCIAL INSTRUMENTS AND OWNERSHIP

Let's take a deeper look at what it means to be an owner of the instruments of the Joint Capital Compound. First, we need to review what we mean by the Instruments of the Joint Capital Compound. We first mentioned this factor of the production and services process in Chapter Two. Allow us to restate:

> All institutions (social, political or religious, public or private); gold, silver, precious gems; all natural resources, tools, machines, and all developed technology are instruments of the Personal and Joint Capital Compounds. When we use the term "Joint Capital Compound" we are simply referring to a collection of all individual Capital Compounds working together. This could be the total group of people represented in a company (sometimes referred to

as the company "family") or it could be the joint efforts
of a city, county, state or the nation.

You will recall that this joint effort has a synergistic byproduct that
provides new growth and new wealth that could not have been
accomplished by any of the individuals working singly. For our
purposes, those who own shares of equity in the instruments of
production of a Joint Capital Compound are owners of the company
or institution in question, including all of its tools, equipment, facilities,
business contacts, and technologies.

RISK FROM A PERSPECTIVE OF OWNERSHIP

Now what does it mean to be an owner of these instruments? In
the pure state of total financial independence, it means that you
would receive a regular income from the profits that were generated
by them. Your medical benefits are most likely purchased outright as
an individual plan, or perhaps a small group plan. Unemployment
insurance is unnecessary, for you have no employment as such—but
you do have an income.

Your retirement program consists of the diversity of your personal
investments. It is not a factor of the future only, as it is in our present
day. You are "retired" inasmuch as you have chosen *not* to work at a job
that would just give you wage compensation. You might enjoy many
different activities, any one of which could require great amounts of
energy and effort. You would probably not refer to it as a job, but rather
as your latest "project," something you do because you like to do it,
not because you have to do it. It's obvious that the American worker
is not even close to a pure state of total financial independence.

Perhaps this fact will help us come to a better appreciation of the
true meaning of the terms haves and have-nots. One might easily
believe that these terms refer to personal possessions such as houses,
automobiles, jewelry, recreational vehicles, home computers, vacations,
and cash in the bank. But in reality these comforts of life are merely a
byproduct of the true quality of "having." To experience the reality of
having means you own equity shares in the instruments of production

of a Joint Capital Compound in sufficient quantity so as to remove the need for wages. Your standard of living is merely a byproduct of such ownership.

A COMBINATION OF WAGES AND EQUITY

If we find ourselves in less than a pure state of total financial independence, we must consider a combination of both wages and distribution of equity shares. In this example we will establish two sources from which to fund our major medical health program, our unemployment insurance, our disability insurance, and our retirement plan: namely, wages and dividends from equity shares. As we have pointed out, the more equity a worker acquires in a given Joint Capital Compound, the less compensation that is required in the form of wages.

As a result of the arrangements for payment of these benefits, the portion of those payments that would be made out of the equity holdings would increase as the worker's equity shares increased. Over time this increase would eventually assume the full level of payment for all of the benefits of the worker involved and no more deductions from wages would be needed. This could be so arranged as to take place long before the worker actually reached a point of receiving total compensation in the form of equity shares.

Under these arrangements a worker who was forced into an unwanted layoff would continue to receive compensation in the form of unemployment income, health coverage, and continued retirement payments indefinitely. In addition, as a worker's level of ownership reached a certain point, there would also be income payments from the profits paid to the shareholders over and above the cost of the benefit package. Ultimately all senior workers would enjoy the advantages of total or near total financial independence. At this point the options regarding the worker's utilization of his layoff time become very broad.

Certainly corporations and businesses need to assume more of a community role than is now practiced. Perhaps workers who have basically been set free from toil through their ownership of equity shares would wish to assume active roles in community affairs on behalf of the corporation. Hence, the workers of the near future could enjoy the privileges of participating in the leisure activities of society without any need to resort to chattel slavery, as the propertied gentlemen of Aristotle's time had to do. Our modern "slaves" would be those that are born of the fruit of our technology.

EARLY CONCEPTS OF RISK AND OWNERSHIP

Although we are surrounded by the fruits of a highly advanced technology, our thoughts and habits relative to compensation are built around the old idea that cash upfront is best. If not from our own personal experiences, then perhaps through the sufferings of others, we need to accept the fact that such an idea is both very *old*, and equally *untrue*. Our reasons for adopting such an idea are even antiquated, intimately linked to the age of the Industrial Revolution, and to our acquiescence to Marx's Labor Theory of Value, and its resultant laborism.

Our earliest roots are quite the opposite. Prior to the time of the Industrial Revolution, the far greater majority of all compensation was made from the direct profits of the world's far-flung agricultural activities. The goal of every landless peasant for centuries was to have a plot of ground that he could call his own. What for? Certainly not so he could go to the center of his fields every Friday afternoon and pick up his weekly cash payment for his time invested!

The land was everything, and one had to invest time, energy, valuable seeds from the previous crops, and faith—faith that after all the work was done, Providence would bless it with rain and sun in the right mixture to bring forth a bountiful harvest. The profits could not be paid out in advance as there were none. No one even knew if there would be any, but the batting average of Mother Nature is so good that everyone wanted in on the act. Remember, this was so despite the droughts and pestilences that often befell the farmers. In short, working the land for oneself was infinitely preferable to working it for someone else—even considering the risk!

In Chapter Two, we mentioned that it was land that afforded man his first real opportunity to utilize the Capital Compound to consistently create new wealth. We also mentioned the impressions that were recorded by Alexander Hamilton in the *Report on Manufactures,* issued by the Secretary of the Treasury in 1791:

> "That in the productions of soil, nature co-operates with man; and that the effect of their joint labor must be greater than that of the labor of man alone."

Therefore we can appreciate the fact that such early leaders were able to recognize the Capital Compound synergism in relation to farming. Even without understanding exactly why, everyone knew that one's best opportunity was in being an owner of land, not a hired hand to work the land.

OWNERSHIP AS A "BIRTHRIGHT"

Wage earners of olden days were far less than second-class citizens. They were equated with the foolish, such as Esau, the Biblical character who sold his birthright. The birthright was a privilege of the firstborn son. This meant that the firstborn son had priority over all the other sons and daughters of the family to inherit the wealth and assets of the family. The greatest portion of all the lands, money, and animals would be given to the firstborn son. It was up to him to carry on the name and honor of the family, and to manage the family wealth for the benefit of all the rest. Esau, in a moment of weak judgment, was tempted to trade all of that for a bowl of soup. In Genesis 25, the writer tells the story in simple, but forceful language:

> Once when Jacob [Esau's brother] was cooking some stew, Esau came in from the open country, famished. He said to Jacob, "Quick, let me have some of that red stew! I'm famished!" (That is why he was also called Edom, which means Red.)

> Jacob replied, "First sell me your birthright."

"Look, I am about to die," Esau said. "What good is the birthright to me?"

But Jacob said, "Swear to me first." So he swore an oath to him, selling his birthright to Jacob.

Then Jacob gave Esau some bread and some lentil stew. He ate and drank, and then got up and left.

So Esau despised his birthright.

Are we not guilty, in a sense, of despising our own birthright by trading the fruits of our Capital Compound for a mere wage?

WE ARE RESPONSIBLE FOR OUR FUTURE

There is no profit in placing blame on the mistakes and shortsightedness of our forefathers during the Industrial Revolution. For one thing, we might not have done any better had we been faced with the same apparently insurmountable odds. Yet, it is our firm belief that after age thirty, the child himself is responsible for the new wrinkles of age and worry that appear in his face. We can only blame our forefathers up to a point. Now is *our* time to face reality, and the challenges that the future holds. What we do with these challenges, and whatever successes or failures result from our decisions, is our responsibility and not that of our forefathers!

THE AMERICAN DREAM—A MYTH OF LABORISM

Wage earners in America have become known as the middle class, and even though many workers would not qualify financially as truly middle class, most consider themselves to be so. When hope sprang up in the hearts of the average American, the great American Dream was born. But we must not lose sight of that on which the hope was based.

Prior to the Industrial Revolution it was based on ownership of property and capital instruments. However, since the time of the Industrial

Revolution it has been based on the strength of laborism. Today we hear rumors that the American Dream is dead, and that the average American can no longer expect to have the same standard of living, or the same quality of future, that used to be considered his right.

Nevertheless, this American Dream spread to all of the politically free countries of the world, and the desire to achieve that dream became the standard for the free world. It's hard for many Americans to believe, but the standard of living in several Western countries is now reported to be higher than that of the average American: Switzerland and Germany, for example. Now, however, the same reversals that we have experienced are starting to happen in those countries as well. The inability of laborism to continue its upward spiral of continually higher wages and higher standards of living for the majority is becoming evident for all to see. In fact, even the demands for "job security" at the same wage level which is now being promoted by organized labor will only result in accelerating the loss of jobs through technology. Today, England, all of Europe, Russia, Italy, Canada and our beloved America are all suffering dramatic negative effects of the fallout from laborism.

For the modern American worker and the workers of other countries of the world, the pursuit of the dream draws the individual farther and farther away from the basic relationship with the land. So we are removed from the laboratory of reality, unable to see the basic lessons of the productivity of our total Capital Compound that living close to the land brings. We no longer identify with the elements of time and of risk in the basic production of wealth. We have become a cozy-conscious culture with little or no understanding of the true source of all wealth. Our recent losses and struggles mean that, to survive financially, we are a much less cozy-conscious culture today!

We live in a country that is known as a Capitalist country. Indeed there are hundreds of thousands of fortunate Americans who are true Capitalists and who know the benefits of ownership, but they are a culture apart from the great majority. We are truly a *divided society.*

There are the few, less than four percent, who can say that they receive their total income from ownership of their Capital Compound and the instruments of production. These are the true Capitalists. The rest range from those who can honestly claim to be partially Capitalists, down to those who are so far removed from such a reality that there is little understanding of the word!

OUR PERSONAL EXPERIENCE

My wife, Marilyn, and I own and operate a small accounting company that specializes in small businesses. Our clients range in size from those companies with only three to five employees to those with as many as sixty. We have been blessed with a core of client companies that have all had much better than average management. This is evidenced by the fact that most of our clients have been with us for over ten years.

For several years we did not seek or acquire any new clients. In July of 2009 we lost a very large client which had become over thirty percent of our gross income. At the moment it seemed impossible to replace that much of our gross revenue in a short time. Seven months later, with the addition of seven new clients, we are back where we were in July. If we had been dependent upon wages instead of our full Capital Compound we could not have recovered so fast.

Our Company, Hatfield & Company, Inc., is small. We have no employees but we have expanded technology that makes us effective, productive, and mobile. Our annual income from these efforts is not enormous but continues to be over twice the national average.

Like so many small business owners, we do not think of retirement. We think of ways to keep working, being more efficient, more productive, and more mobile. We have twenty-four grandchildren spread from coast to coast. Our mobility allows us to travel and to visit them all once or twice annually.

Naturally, we are taking steps to strengthen our financial independence as we progress in age. We know that we cannot always be as effective and productive as we are now. The latest development in America's economy presents a more challenging environment in which to grow, but even this is not strong enough to cancel the power of our personal Capital Compound.

There is, however, a point at which no small business can survive. The negative impact of Obama's Progressive Socialist movement is capable of destroying the country's small business enterprise. Currently, America's small business sector boasts the world's largest and most efficient enterprise system. We create more new jobs. We create more Gross National Product—we are the hope of America's financial future!

We are also the nemesis of Obama's Progressive Socialist movement. Without us there is nothing to stand in his way to take over the entire economy. Sadly, the government cannot create new jobs, but they have the power to destroy jobs. Only the arrogance of Progressive Socialist elitists could simultaneously seek economic power and at the same time destroy the very essence of its source.

As Glenn Beck said in his CPAC Keynote speech; "Our future is not cast in stone. It does not have to be this way!"

THE DIVIDING LINE

The most critical division of our society is not along geographical lines or color lines. The most critical division is clearly one of ownership. On one hand we have the owners of the capital instruments of production (the expropriated production of our intangible elements and their synergistic rewards) who are the haves. On the other hand we have the far greater majority that receives only partial compensation for their total Capital Compound contribution—they are the have nots. In which class do you find yourself?

WHAT DOES NOT WORK

Before moving into our section on specific solutions we must come to grips with what does not work, cannot be made to work, and cannot be fixed. This something is a very good source for addressing some of our great needs as a country. It is designed to function in the way that it does because that is its primary function. All other uses of this entity have proven to be very inappropriate and disastrous! Hanging on to the notion that this entity can serve us well beyond what it was originally created to do is like attempting to swim the ocean with a ton of steel on our backs.

Reagan said it best: "The solution to our problems is not the government; government is the problem!" Government was originally created and conceived to provide national protection from our enemies abroad, and to supervise in a larger sense the national impact of the security provided internally through each state. The government is perfect for the execution of its original purpose and design. To perform its original purpose and design, our government—headed up by our Commander-in-chief, our elected president—seeks out and identifies our external enemies. When these are identified it must decide if the enemies in question only need diplomatic adjustments or do we need to declare war.

War is a terrible thing, but war is what our government was created to do and to do it successfully every time it is called upon to do it. Government, by its very design and the nature of its most inherent abilities, is best suited to seek out and destroy. Government was never intended to be a nursemaid, a manager of civilian lives and personal affairs. This is the very thing that drove our colonial leaders out of the old country: centralized, controlling, over-bearing, dictatorial governments of all stripes.

When these United States allowed our beloved Constitution to be interpreted by Progressive Socialists to mean it had the power and obligation to be a nanny government because of the phrase "general welfare of the people" we gave birth to the monster that we now have. Our framers were referring to the government's role as our major element of international protection, not for our need of a nanny. If our framers had intended these words to mean a nanny state, they would have repented of the revolution and humbly returned the country back to mother England, our nanny government of the day!

The aberration of what we were given, compared to what we now have, was never in the minds of the framers of the Constitution. We were sold a bill of goods by FDR and the rest of his Progressive Socialist ilk and now we have this full-grown monster on our hands.

The monster is doing exactly what the monster—i.e. the government— was created to do. It is going about destroying everything it touches. This is great, when what it touches are our enemies and those that would destroy our nation. When you turn an eight hundred pound gorilla lose in a canary cage, you had better get prepared to bury a lot of dead canaries!

Check the record. What institution has the government created that it has not destroyed, except the military? Government is a destructive element by nature, and woe to those who think it can be tamed and turned into a house pet. Government is so dangerous that it has a very difficult time even to manage something as closely associated with the military as the FBI, the CIA, and the NSA. It is no wonder that government cannot manage Social Security, welfare, IRS, Medicare, Medicaid, Fanny Mae, Freddie Mac, or the Federal Reserve System, not to mention the hundreds of other institutions that it has created or claimed as its own! Are we insane or what?

WE MUST PRESS ON

"Those who expect to reap the blessings of freedom must, like men, undergo the fatigue of supporting it." — Thomas Paine

Following is a speech given by J. D. Hayworth on March 22, 2010, a day after the notorious Health Care Bill was voted in.

With the roar of thousands of protesters screaming "NO" and "KILL THE BILL" outside the United States Capitol, Nancy Pelosi and House Democrats defied the will of the American people voting 219-212 in favor of Barack Obama's health care reform legislation.

Let's be honest: Freedom and Liberty took a punch to the gut, and it really does sting.

We cannot afford to throw a pity party. We must meet this challenge with fortitude and determination. Now, more than ever, Arizonans and Americans must show their true spirit and strength, not wallowing in failure but showing persistence and courage to press on and institute positive change.

The darkest hours in our history are met with the great American spirit. Each generation has had its fight and opportunity to change the course of American history. This is ours!

It is time for the American people to band together and fight for the great freedoms that this country represents. It is time to show the true spirit and power each and every American has inside of us.

The fight over health care has only begun, and although we have fallen, we will get up. And we will be heard. And we will continue to fight. Even if we have to wait until the polls open on Election Day.

Defend Freedom,

J.D. Hayworth

SUMMARY

Once again, let us soak in the words of President Reagan: "The solution to our problems is not the government; government is the problem!"

In previous chapters we have seen clearly that the Progressive Socialists are enemies of our Constitution. We naturally think of the many ways that are available to protect ourselves when we perceive any danger nearby. In the case of our enemies and the dangers that we face, there is no insurance policy that will meet the challenge. Our only protection from this voracious enemy is our awareness, our valor, our courage, our faith that God is with us, and our commitment to never give in.

With this attitude we will excite the intangibles of our Personal and Joint Capital Compounds. With our creative initiative, spiritual inspiration, and synergism we can expect that, by the grace and providence of God, He will propel us forward and equip us with all that we need to prevail!

Chapter Eleven

Ideas Have Consequences

"Dream no small dreams for they have no power
To move the hearts of men"-
Johann Wolfgang von Goethe

Insanity is often described as a consistent attempt to succeed in a given goal
by doing the same thing over and over again even though it never works.
Great, now we know one symptom of insanity. This doesn't really help us
unless we are willing to make an informed *change to our approach. Power*
to change is great, but to have the power to change without knowing where
to apply it and what direction to go in to effect the change that you want
is useless. How many times have American conservatives had the majority
power in the House of Representatives, the Senate, and the White House
only to then demonstrate that they did not know how to apply it or in what
direction they needed to go to effect the change we must have to recapture
completely our beloved republic?

UNDERSTANDING THE QUESTIONS

If we don't understand our own problem enough to know how serious it
really is, then we probably don't know enough to ask the most important
questions. If we don't know the questions to ask, how would we know

that the answer given would actually solve the problem? The following are three key questions:

What are the creative possibilities for reducing the risks associated with the creation of new wealth which we expect from all three elements of our personal Capital Compound?

How does the Joint Capital Compound provide the heavy lifting that Progressive Socialist elitists say only the federal government can do?

What constitutes "unemployment" in a society where the primary form of distribution is through ownership in the instruments of production within our total Capital Compound?

The purpose behind this book is not simply to join the ranks of the many critics and pundits of the Obama administration. We applaud and deeply appreciate and respect all of those who have brought forth many very important points to educate all of us on the inner workings of this arrogant and misguided administration. However, knowing enough to criticize—even to do so very brilliantly—does not automatically provide the wisdom for not repeating the same mistakes again. As we have said before, our predecessors have been here before several times over the last hundred years. Yet, here we are once again, only now the decay and corruption is so deep that we can't escape the odor of death all around us.

Solutions require ideas and ideas have consequences! Our ideas as presented here in this book come from our many years of experience and research into the dynamics of the principles of true Capitalism and how our constitution has opened the door for us to apply these principles. We want you, America, to consider the wisdom of what we present and the proven dynamics of the solutions. America, you are very blessed with the ability and the common sense to understand what we are going to propose. You read, you consider, and then you decide.

OUR EARLY EXPERIENCES AS SMALL BUSINESS OWNERS

One of our earlier companies was a financial planning service located in Santa Rosa, California. This gave us a great experience in the field of stocks, bonds, limited partnerships, and insurance. In this company we did have some employees and our profits were larger. Over the years as we gained more experience, we began to specialize in capitalization of medium-sized startup companies.

To do this service we were required to hold several different licenses. We needed several broker dealer licenses:

Series 65   Liability and law

Series 7   Stocks, bonds, mutual funds, limited partnerships

Series 24   OSJ or Office of Securities Jurisdiction

In addition to these we needed insurance licenses in life and health coverage. Although not required we also had a real estate license.

During our twelve years in this industry we were introduced to many financial tools and the many details associated with the process of creating new capital through investors. We were very successful in our efforts to serve the many new startup companies that came to us for service. In this world of finance one gets to see the dynamics of wealth creation from a very different perspective than working for a wage. This is the heart of Capitalism with all its warts, blessings, and cursing.

Capitalism is the short answer to the three questions above. The specific steps and tools of Capitalism that are used to fully answer these questions are a little more complicated, however.

Capitalism at its core is the economic expression of the intangible elements within our personal Capital Compound.

PERSONAL CAPITAL COMPOUND

1. Our personal labor - tangible element

2. Creative Initiative - intangible element

3. Spiritual inspiration - intangible element

4. Synergism - the unexpected compounding by-product of the whole

FIRST QUESTION

What are the creative possibilities for reducing the risks associated with the creation of new wealth which we expect from all three elements of our Personal Capital Compound?

The answer to the first question lies primarily in the creative development of new applications of existing financial instruments available in our capital markets. There is a very little known capital instrument that has provided historic success in the field of investment real estate. This capital instrument is called a REIT, or Real Estate Investment Trust. This capital instrument is less than forty years old and has completely revolutionized a sizable segment of the real estate industry, especially in the area of commercial real estate. Our current recession, which has devastated personal ownership of real estate, has had only a marginal negative impact on the REIT industry. At the last report, all REITs are still paying quarterly dividends.

It is especially noteworthy that the modern REIT is now the preferred method for raising and managing the capital required to own and control large high-end property assets. In addition, it is equally noteworthy that the private REIT now exceeds, by a considerable margin, the publicly traded REIT. This latter fact is due primarily to the high maintenance costs associated with the regulatory requirements necessary to qualify and operate as a publicly traded REIT.

To qualify as a REIT, ninety-five percent of all net profits must be distributed to the investors: these cost savings are automatically passed on to the shareholders through the quarterly dividend payments required of any REIT.

UNDERSTANDING REIT FORMATION AND OPERATIONS

The following is an excerpt from reit.com detailing the formation and operation of any REIT.

Forming and Operating a
Real Estate Investment Trust

The following generally summarizes some of the basic tax law requirements applicable to **REITs**. These rules are complex, and the following is only a general summary. In order to qualify as a **REIT**, an entity must meet a number of organizational, operational, distribution, and compliance requirements. If the **REIT** satisfies these requirements, it will be entitled to deduct any dividends paid from its taxable income. A **REIT** that distributes 100% of its taxable income therefore will have no federal income tax liability. Although state tax laws relating to **REITs** vary, most states with an income-based tax regime follow the federal law and permit a **REIT** a "dividends paid deduction.

1. **Organizational**

A REIT must be formed in one of the 50 states or District of Columbia as an entity taxable for federal purposes as a corporation. It must be governed by directors or trustees, and its shares must be transferable. Beginning with its second taxable year, a REIT must meet two ownership tests: it must have at least 100 different shareholders (the "100 Shareholder Test"), and 5 or fewer individuals cannot own more than 50% of the value of the REIT's stock during the last half of its taxable year (the "5/50 Test"). These ownership requirements generally mean that the REIT structure is not a good choice for a closely held family business. A number of "look through" rules currently apply when determining whether the REIT meets the 5/50 test.

In an attempt to ensure compliance with these tests, most REITs include percentage ownership limitations in their organizational documents. For example, many REITs do not permit any one shareholder to own more than at most 9.9% of a REIT's stock without a waiver by the REIT's board of directors. Because of the need to have 100 shareholders and the complexity of both of these tests, general legal, and tax and securities law advice are strongly recommended prior to beginning the process of forming a REIT.

2. Operational

The REIT must satisfy two annual income tests and a number of quarterly asset tests that are designed to ensure that the majority of the REIT's income and assets are derived from real estate sources. Annually, at least 75% of the REIT's gross income must be from real estate-related income such as rents from real property and interest on obligations secured by mortgages on real property. Additionally, 95% of the REIT's gross income must be from the above-listed 42 sources, but can also include other passive forms of income such as dividends and interest from non-real estate sources (like bank deposit interest). As a result of these rules, no more than 5% of a REIT's income can be from non-qualifying sources, such as from service fees or a non-real estate business. A REIT can own up to 100% of the stock of a "Taxable REIT Subsidiary" ("TRS"), a corporation with which a REIT makes a joint election that can earn such income. Quarterly, at least 75% of a REIT's assets must consist of real estate assets such as real property or loans secured by real property. Although a REIT can own up to 100% of a TRS, a REIT cannot own, directly or indirectly, more than 10% of the voting securities of any corporation other than another REIT, TRS or accredited REIT subsidiary ("QRS"), a wholly owned subsidiary of the REIT whose assets and income are considered owned by the REIT for tax purposes. Nor can a REIT own stock in a corporation (other than a REIT, TRS or QRS) the value of whose stock comprises more than 5% of a REIT's assets. Finally, the value of the stock of all of a REIT's TRSs cannot comprise more than 20% of the value of the REIT's assets.

3. Distribution

In order to qualify as a REIT, generally, the REIT must distribute at least 90% of the sum of its taxable income. To the extent that the REIT retains income, it must pay tax on such income just like any other corporation.

4. Compliance

In order to qualify as a REIT, a company must make a REIT election. The REIT election is made by filing an income tax return on Form 1120-REIT. Because this form is not due until, at the earliest, March 15th following the end of the REIT's last tax year, the REIT does not make its election until after the end of its first year (or part-year) as a REIT. Nevertheless, if it desires to qualify as a REIT for that year, it must meet the various REIT tests during that year (with the exception of the 100 Shareholder Test and the 5/50 Test, both of which must be met beginning with the REIT's second taxable year.) Additionally, the REIT annually must mail letters to its shareholders of record requesting details of beneficial ownership of shares. Significant monetary penalties will apply to a REIT that fails to mail these letters on a timely basis.

Disclaimer

Please note that the discussion set forth above is for informational purposes only and is not intended to constitute legal or tax advice. Because the formation and operation of a REIT involves many complex legal, securities, tax and accounting rules, we strongly advise you to seek professional advice from competent attorneys, accountants, and other advisors prior to beginning the process of forming a REIT. Since we are not providing legal advice through this Web site, you should not rely upon any information contained herein for any purpose without seeking legal and/or tax and accounting advice from a duly licensed attorney or tax practitioner.

COPYRIGHT © 2006 - National Association of Real Estate Investment Trusts®

This financial instrument is only one example of how to create wealth through the investment market and at the same time provide exceptional safeguards for the investors. It is no wonder that the REIT is one of the most popular investment vehicles for many large retirement investment portfolios.

SECOND QUESTION

How does the Joint Capital Compound provide the heavy lifting that Progressive Socialist elitists say only the federal government can do?

The second question is much like the first question, only with an emphasis toward how to create wealth to cover the needs of large numbers of people. Since it is impossible to separate risk protection from asset wealth accumulation, the answers are going to seem very similar. However, for this answer I will use the state of Alaska.

Many people have heard about how Alaska pays their qualified citizens annual dividends from the exploration and sale of oil. What you might not know is the instrument through which they do this, the ESOP or Equity Stock Option Plan.

An ESOP is a financial instrument that is groomed to fit the group in question. It could be a family business that will one day lose its founder and the family wants to assure themselves that the assets will continue to be managed successfully. So, instead of selling the company to an outside buyer, the current family sells the company to their employees through this instrument. In the case of Alaska, it is a little different.

Alaskan state citizens are assumed to be joint owners of all of the oil deposits of the state (in this case we are talking about the oil revenues). Governor Sarah Palin acquired considerable notoriety when she was able to increase the profit portion for the state over what the oil companies had fraudulently bargained for in previous years. Through this recognized ownership, each citizen qualifies for a percentage of the income called a dividend. You will remember that some state officials are now serving time for their part in the fraud.

Question one and question two deal with certain areas of our Capital Compound. Question one is more concerned with creating new wealth and low risk. Question two is more concerned about distribution of existing wealth to a broad number of individuals. Both aspects embrace different private financial instruments of Capitalism, not government. There are no actual needs or opportunities that require new capital that private capital cannot meet without the need to involve government.

Through the influence of those who hate Capitalism (namely, the Progressive Socialists), government has been sold to America as an alternative to Capitalism. When we accept the arguments of the Progressive Socialists as true then, by default, we are accepting Communism.

Is it possible to use financial instruments of Capitalism to bring ownership of capital instruments to every American? Absolutely! Our only obstacle is ourselves. So long as we, the American people, accept labor and government as viable and preferred alternatives to Capitalism, we will never fight for our opportunities to own these instruments. We must admit our failures of the past and re-educate ourselves to the reality of true Capitalism, which is the *ownership of instruments of production* for everyone and not just the few. Ironically, the few have not forced us to take wages instead of capital, we ourselves have demanded it!

UNEMPLOYMENT REFOCUSED

THIRD QUESTION

What constitutes "unemployment" in a society where the primary form of distribution is through ownership in the instruments of production within our total Personal and Joint Capital Compounds?

In order to answer our third question, we must recognize that unemployment and laborism are part and parcel of the same problem. So long as we support the concept that the primary form of distribution should be via a wage, we will have boom and bust cycles of employment and unemployment, with the unemployment becoming more serious as technology increases.

In stark contrast to this, the Capital Compound Theory of Value demands recognition of all three elements of our Capital Compound, and a just distribution of equity shares to match. Our current labor element is grossly overvalued, while at the same time little or no recognition is given for the two most important elements, our creative initiative and spiritual inspiration. Whereas at first it will appear for many owners of companies that it is unfair to them to have to consider a concept of distribution that includes giving up a percentage of the ownership, the truth is that it is really to their advantage.

Certain major companies (and even a few labor leaders) are now beginning to see these advantages and are moving cautiously in that direction. Our constitution is our foundation for personal property rights. Such recognition of ownership allows control, control gives rise to the possibility of security, and the hope of security encourages involvement. Consequently, it is reasonable for us to be more protective and diligent with what we consider to belong to us, as opposed to that which belongs to someone else!

Insisting on a primary distribution concept of labor and wages guarantees a continuation of up and down cycles in unemployment. However, if a worker should receive enough equity shares over a period of time to be able to live well without the toil of labor, such a person would not be employed in the accepted sense, but neither would he be unemployed!

In fact, it is just this possibility of liberated workers that holds one of our greatest opportunities for community and civic action. These individuals might very well accept as part of their working agreements that, as their equity shares begin to provide certain levels of their overall income, they would accept more community involvement under company direction.

In other words, they would have a divided workweek. Part of their work would be directly related to the products or services of the company and the other part would be directed to the company's community and civic interests. Such an expanded use of our creative efforts would make for stronger and more successful communities and at the same time reduce the tax burden and the role of government in those areas.

As workers become more sensitive to the fact that there has been an expropriation of their intangible elements, they will demand more distribution in the form of equity shares and will accept less in the form of wages. The American worker's willingness to overcome his/her uninformed state will determine how fast we overcome the phenomenon of unemployment.

Once a grassroots movement is underway to reverse our laboristic concepts, unemployment as we have known it will, in time, become a thing of the past. Some people will still suffer from periods of temporary unemployment while the country adjusts from laborism to a just distribution of our total Capital Compound. Nevertheless, reductions of unemployment could be very rapid, and worker morale and faith in the future would improve dramatically.

With the internet as our most dynamic national and international communication tool, TV, radio, and newspapers no longer hold a monopoly on getting out the word. Concerned workers and individuals could bring this issue to the forefront very quickly. The whole transition could conceivably take less than ten years.

But the American workforce is the key. By understanding their true values and how those values should be compensated, they will no longer continue to "bargain for a penny," leaving the bulk of their wealth in the hands of others!

CREATING NEW FINANCIAL INSTRUMENTS

We have limited our discussion of capital financial instruments to the ESOP and the REIT. We have done this because they offer us the most dynamic models of creative postmodern financial structuring.

These two examples are only the tip of the iceberg! There are thousands of practical applications for these instruments and thousands more awaiting the results of our creative initiative and spiritual inspiration elements of wealth creation. In designing these future financial instruments some legislation will be necessary to get their approval and acceptance within our current system of understanding. Some laws will be needed to rectify old mistakes and open new doors. Other laws will be needed to protect the ownership rights of those who choose to use a given financial instrument to best facilitate their wealth creating activities. One thing is for sure: there are enemies to such financial instruments. The two most significant are union leaders and Progressive Socialists.

Despite the fact that the unions will present great opposition to the transfer from laborism to Capitalism, there is little they can do to prevent the change which starts first in privately-owned companies that are not associated with unions. Remember, seventy percent of America's National Gross Product comes from businesses with less than one hundred employees. Union representation of profit generating workers in America is less than six percent—the rest of their representation is for government workers of all ranks and types. They have one hundred percent of this labor force, but it is an impotent labor force in that it does not create one thin dime of profit.

Politically, the Progressive Socialists will fight the change at every opportunity. They will impugn motives. They will attempt to destroy every public official who works for this change from laborism to true Capitalism. They will declare all who seek this change to be the enemies of the common men and women. They will accuse us of being cruel and heartless and that what we propose will impoverish more people than ever in our history. Progressives will enlist the help of the union leaders to bring personal threats to their memberships everywhere if they join in the struggle to move from laborism to pure Capitalism.

Progressive Socialists hate Capitalism and pure Capitalism they hate even more. Pure Capitalism is the death of all that they stand for. Without the excuse to "help" the masses who can't help themselves, there is no wind in their Socialist ship of state. Their ship of state needs lots of victims to justify their greed and appetite for power. These victims are the wind that propels their ship forward. A successful return to true Capitalism, such as our colonists enjoyed, is the worst thing that a Progressive Socialist can imagine. For them it is the same as a new sunrise for a vampire—death!

In stark contrast to their Socialist ship of state, our Capitalist republic ship of state has strong steady winds generated by the force of equity ownership and equitable distribution, of the wealth creating power of our Personal and Joint Capital Compounds. We sail straight and sure into the rising sun without fear, for we know that we sail into *life*. We know our true strength is in the grace and providence of God. We have our inalienable rights from Him alone.

SUMMARY

At the beginning of this chapter we spoke of insanity as a consistent attempt to succeed in a given goal by doing the same thing over and over again even thought it never works.

What we must now consider is that the first example of insanity does not exactly describe our symptoms and our situation as clearly as it should. In light of all that we have shared (which is true and undeniable) we must seriously consider that true insanity is when we know who the enemy is, when we know how to take his power away, when we know how to correct our mistakes, when we know what steps and what directions to take, and then we wimp out, turn our backs on our blessed destiny, and allow the enemy to destroy us and our beloved America. This, my fellow Americans, is *true insanity!*

Arise America, rebuild your God-given capitalist foundations!

Chapter Twelve

Refocusing Taxes

The real inconvenient truth is the fact that taxes are a necessary element of successful government at all levels, local, state and federal. What is more important is the fact that most of our taxes are not necessary in an economy that is centered and focused on the Compound Theory of Value. Taxes are necessary, but unnecessary taxes are an abuse of power and a testimony to the greed, arrogance, and ignorance of the leaders who propose them and who sustain the Progressive laborism that creates the need for them! The most alarming inconvenient truth in all of this is that "we the people" control all taxation through our voting habits and our acquiescence to false propaganda!

A BAD WORD

Taxes in any language are a bad word, and especially for anyone who feels that they are paying more than they should. Any discussion of the creation of new wealth, and the distribution of that wealth within a society, cannot avoid the subject of taxes. It is one thing to achieve a measure of wealth through equitable means distribution and productive relationships, and it is an entirely different matter to be able to keep it.

Proverbs 21:20 says "In the house of the wise are stores of choice food and oil, but a foolish man devours all he has." One of the fastest and sometimes least advisable ways a person can find to spend, or devour, their money is through taxes. Some taxes could even be deemed to be foolish at best and unconstitutional at worst.

TAXES HAVE LIMITS

The early American colonists were subjected to many different kinds of taxes by England. When the Stamp Acts were passed by the English Parliament in 1765, the colonists were required to purchase stamps for almost every act of commerce. They cried out, "Taxation without representation is tyranny!" England would not compromise, and subsequently the American Revolution was born.

Taxes have a point of balance. If government charges no tax, they will collect no revenue outside of volunteers. Conversely, if government charges a one hundred percent tax, they will collect no tax except from those who are willing to work for nothing. Neither extreme is practical, and realistic levels of taxes must be sought that balance what the government thinks it needs without ever asking for so much as to discourage its citizens from aggressive success.

Even with the best of balances almost everyone complains about taxes, and yet only a few claim to know anything about them, other than the strongly held belief that you have to pay them. I'm sure everyone has heard that there are only two things in life that one has to do: die and pay taxes. Gerontologists are working on the first one, and economists have given up on the second. One frustrated taxpayer put it like this: "By the time I found out what the answers were, someone had changed the questions."

TAXES IN REVIEW

In order to refocus our perspective of taxes and how they relate to the creation of new wealth and its equitable distribution, we first need to put into focus our basic understanding of taxes as we now use and apply them. To do this we will examine these five questions:

1. What are taxes?

2. What are the uses of taxes?

3. What are the different taxes we pay?

4. Who or what controls taxes?

5. Who pays taxes?

WHAT ARE TAXES?

Historically taxes have been accepted as the correct and legal method by which governments or ruling bodies acquire the revenues that they need to carry out their duties and functions. From the earliest recorded histories of ancient civilizations to the present, some form of taxation has been levied upon the people by their respective governments.

Tax payments in the beginning were in the form of goods and services such as animals, grains, and other commodities. Personal services would have to be rendered in cases where the person being taxed had no other way to make payment. Later, as currency such as gold and silver came into use, this method of payment became increasingly popular. Long before the Industrial Revolution and the adoption of distributing wealth primarily through wages, taxes have customarily been collected in the form of currency only.

WHAT ARE THE USES OF TAXES?

At first glance it would seem that we had answered this question with the statement that taxes are for the purpose of paying the cost of government. For all Americans who pay taxes, however, for those who pay little and those who carry the largest portion of all, such an answer is too simplistic.

In examining the purposes of our taxes we have to consider the motivations and goals of our local, state, and federal governments. These goals vary from community to community as they do from state to state. So we have goals and objectives that local and regional governments want to meet with the available taxes. At the same time the federal government has national and international projects and goals that it wants to accomplish.

Government is often viewed by the constituency in general, and working people in particular, as something inhuman. Certainly the mistakes of government can be so ludicrous as to appear to come from sources other than human. The truth, however, is that government in a democratic society is only a body of people, hopefully serving the best interests of those who have elected or appointed them to office. Consequently governments, like individuals, are influenced by the various philosophies, ideas, and concepts they are exposed to. When they are having problems and running out of ideas, they can be influenced by new ideas that appear to hold out hope for overcoming their particular problem of the moment.

John Maynard Keynes (1883-1946) proposed some new ideas for all non-Communist governments emerging from the chaos of the First World War, and later for those same governments as they were reeling under the devastating blows of the Great Depression in the 1930s.

Keynes was born in Cambridge, England, and studied at Cambridge University. He served in the British Treasury from 1915 to 1919, and wrote a book called *Consequences of the Peace* (1919), in which he attacked the reparations (payments) which the Allies demanded from the defeated Central Powers, and predicted the breakdown of the Versailles peace settlement. Keynes continued to write a series of articles over the next few years that contributed greatly to major changes in attitude by non-Communist governments throughout the world. The most important of these were *A Tract on Monetary Reform* (1923), *The End of Laissez Faire* (1926), and *A Treatise on Money* (1930).

Before Keynes' time, non-Communist governments had followed the economic theory of Laissez Faire. This theory maintains that government should not interfere in economic affairs, either to help or hinder. The ruling owners of capital instruments had been successful in convincing governments that the economy would be better off without their intervention. Meanwhile labor leaders, having adopted the concepts of Marx and others that the only source of wealth was labor, struggled for full employment while the owners of the capital instruments only wanted necessary employment, and this as inexpensively as possible.

These conflicting goals of workers and factory owners coupled with constant rapid growth of technology created an atmosphere of cutthroat competition and a tendency to form large monopolies. As a result there was a continual cycle of boom and bust, with recurring massive layoffs and a gradual buildup to the next bust

Keynes attacked these government policies of inaction in general and the English government in particular. His earlier works led to the climax of his economic literary achievements in his book *General Theory of Employment, Interest, and Money* (1936). This work ranks as one of the most influential books on economics ever published in the free world. It revolutionized economic theory and policy, and plays a strong influence in the economic policies of most non-Communist nations today, including America. Up until the time of the Great Depression, Keynes' theories were not widely accepted. Yet, when the depression struck in full force, government leaders in all of the major Western countries began to question the theory of Laissez Faire and to take his ideas more seriously.

The basis of Keynesian economics is simple. It holds the view that the level of economic activity depends on the total spending of consumers, business, and government. If for some reason business feels that its future is negative, it will withhold investment spending, causing a series of reductions in total spending. When this happens, the economy can move into a depression and stay there.

To avoid the boom and bust cycles and to guard against the danger of depressions, Keynes urged increased government spending and "easy money" (lower interest rates and making more money available for loans). Keynes believed that such government action would increase the available money in the hands of the consumer (the workers), which would thereby increase demand, encourage investment, and thus increase employment. His analysis further showed that high levels of demand were essential for both full employment and economic growth.

Actually, this concept would have fallen on deaf ears, and might not even have entered Keynes' mind, if it had not been for the expanded development of laborism as a modification of Marx's Labor Theory of Value. Keynes was simply offering the most logical approach for dealing with the reality of the moment, as it had been developed by labor leaders and workers. Keynes was not a card-carrying Communist, but he was a friendly sympathizer. He simply looked at the "reality" of his times, the era of laborism as the primary means for the distribution of wealth.

No one—not Keynes, nor anyone else—ever asked the question: "What about the other elements of value, what about the intangible elements which create the most value in equity?" They did not ask because they did not know to ask!

ENTER PRISDENT FRANKLIN DELANO ROOSEVELT

Roosevelt, (1882-1945), the first American President to put into practice the concepts of Keynesian economics, was a devoted Progressive Socialist! Roosevelt first took office on March 4, 1933. He faced the unenviable task of leading America out of the worst depression in history. Unemployment was over twenty-five percent, and bread lines stretched for blocks. The future had never looked as bleak as it did in 1933.

Roosevelt is the most influential model for Progressivism and all of its proponents, no matter their political alliances. Obama and his advisors are all outspoken followers of Roosevelt, as are many Republicans like John McCain and his supporters. Progressivism is the new face of Socialism and the most visible political descendant of Communist theory in our postmodern times.

Roosevelt's reform program included a wide range of activities which he called the New Deal. The President described it as a *use of authority of government as an organized form of self-help for all classes and groups and sections of our country.* Roosevelt lost no time in putting his plan to work. At the President's request, Congress appropriated $500 million for relief to states and cities through the Federal Emergency Relief Administration. The Civilian Conservation Corps (CCC) operated from 1933 to 1942, and gave work and training to five hundred thousand young men. The CCC made great progress with its programs of flood control, forestry, and soil conservation.

In 1935 the President got approval from Congress for the Works Progress Administration (WPA) to provide work for unemployed persons. The WPA employed an average of 2,000,000 workers annually between 1935 and 1941. You had to have the right politics to get hired, however. Many in our own family who lived at that time were turned down because they were Republicans. So you either kept your mouth shut or you lied.

In the first year of the President's term in office Congress passed laws to protect and aid various sectors of the economy and the country. Included were such measures as laws to protect the investments of persons who buy stocks and bonds, legislation to help the oil and railroad industries, small businessmen, and homeowners. The Social Security Act was passed in 1935 providing unemployment relief and old-age assistance. The National Labor Relations Act was also passed in 1935, giving workers the right to collective bargaining.

Never before in American history had we experienced such a drastic change in the attitude of government concerning its involvement in the country's economic and social problems. Keynesian economics was given practical life and application with a vengeance. This has been the basic policy of government in America, with only minor variations, to the present. Hence, distribution of wealth through laborism and direct government payments to the public—Social Security, unemployment, disability insurance, health insurance, welfare and food stamps—have all supported the laboristic concept of Full Employment directly or indirectly.

We can plainly see that, in order to understand the uses of our taxes, we must know something about the *goals* of our government. If any society wants to accept the philosophy of Marx that labor is the only source of all wealth, and at the same time reject his idea of the state owning all the property and capital instruments, then a mixture of private ownership and laborism is inevitable. This demands that a policy of Full Employment be adopted (as it was officially in 1946) and that we continue to suffer boom and bust cycles. T

he only big changes that over seventy-five years of Keynesian economics have provided are ever-increasing government size and involvement in our personal lives, enormous deficits, and a constant erosion of the value of our money! These practices, coupled with the enormous interest payments the government has to pay to support the deficits incurred, constitute the largest portion of our tax use.

ENTER PRESIDENT OBAMA AND HIS PROGRESSIVE SOCIALISTS

Everything that Roosevelt did Obama and his administration have duplicated in spades. The main distinction between the Roosevelt administration and the Obama administration is the extreme expropriation of private property. Left unchallenged, this trend toward expropriation of private property will only expand and find new rationalization for more and more acquisitions, as we have already seen with automobile industry and the banking industry. Obama is now moving legislation through Congress to take over the Private Student Loan Industry.

FOLLOW THE MONEY

Obama and the Progressive Socialists vote massive amounts of bailout money for various industries, all in the name of "saving" not only the American economy but the world economy. Progressives from all parties join in this rush to "save" America and the world. However, along with the bailout the administration included several egregious demands for control. Once the bank officials realized the big government strings that were attached, they immediately made plans to pay it back.

Amazing, isn't it? Financial institutions that claim that they and their worldwide counterparts are all on the verge of collapse—and then wow!—we don't need the money after all, thank you very much. Then, to add insult to injury, these same institutions not only pay the money back (save one or two), they report some of their greatest financial gains in history. And all of this in less than one year!

Do you feel as raped as I do? These institutions simply read the tea leaves of the Obama ideology of Progressive Socialism and recognized a great opportunity to con the government and the American people. Greed knows no frontier, no party affiliation, no ethnic preferences: it is one of humanity's great sins. This sin festers in the heart of Progressive laborism more than in any other socioeconomic environment!

Without our national submission to the erroneous concept of the labor theory of value and its expression in laborism, there would never have been an atmosphere in which America could have been so deceived. Without our national submission to Progressive Laborism we would not have a man like Barak Obama in the White House, and we would not have the monster of Progressivism hovering over our heads ready to decapitate us. Without our addiction to this wrongheaded laborism, America would be a modernized version of the early colonies. Americans would primarily be true Capitalists, owners of the instruments of production and in control of their own personal Capital Compound.

APPROPRIATE USES OF OUR TAX DOLLARS

There are appropriate uses of our taxes which are a direct benefit to the economy, and to society as a whole, that do not have the same bad side effects as propping up laborism. These are such things as protection of our private property rights, protection of our civil and personal rights, protection of our religious freedom, protection from internal abuses, and protection from terrorists and warring countries.

In retrospect, the most glaring omission within Keynesian theory of economics and other variations of Marx's Labor Theory of Value is not recognizing the expropriation of the two intangible elements of our Capital Compound and its synergistic contribution. This omission is committed largely because of the false conviction that technology is only the product of a few, and that it will never be able to develop a replacement for the worker in the workplace. Based on those assumptions, and considering that the voting constituency in America suffers tremendously from a lack of knowledge and insight concerning the Capital Compound Theory of Value, our government administrations have acted predictably. Both Democrats and Republicans are equally ignorant of the facts.

At least eighty-five percent of Americans are blessed with a measure of faith and common sense. I have great confidence in America to wake up and to recognize from where true wealth comes. Americans will wake up and will embrace their total Personal Capital Compound. Americans will demand their portion of the profits and equities created by the intangibles of their personal Capital Compound. Americans will recreate once again their original God-given foundations of Capitalism, the ownership of equity and private property for all.

TECHNOLOGY AND LABORISM

Concerning the possibility of technology having the power to replace the worker in the workplace, a quote from one of our present day scientists in the high tech field might be appropriate to mention. This quote was reprinted in *Applied Artificial Intelligence Reporter* published by the University of Miami Intelligent Computer Systems Research Institute. The quote is from Alan Turing, who says:

> It is customary. . . to offer a grain of comfort, in the form of a statement that some peculiarly human characteristic could never be imitated by a machine. I cannot offer any such comfort for I believe that no such bounds can be set.

It never ceases to amaze me how we live and work daily with the marvels of technology, and yet we never seem to connect the dots. Technology washes our dishes, our clothes, our cars and trucks. How many workers would we need to hire to replace just this small use of practical technology? We take as commonplace such things as airliners, bullet trains, and large breweries run by three people. We pull out cell phones and talk to all of our personal and business contacts. We work on the internet and visit locations all around the world in just minutes. We watch news from around the world as it happens and we watch as the Orbiter launches for space and again when it returns to land. Consider how many people would have been necessary to do all of these things, and how they would have had to do it, without the use of technology? The truth is that most could never be done without technology no matter the amount of labor employed to do it. What does this us?

Technology certainly exists and is the direct expression of the intangible elements of both the Personal Capital Compound and the Joint Capital Compound. Technology is the consistent witness of the efficiency and power of these elements and the synergism they generate.

We all witness this in our personal lives. Why then are we so blind as to not understand the impossibility of labor to ever outpace or out produce the power of the total Capital Compound and the Joint Capital Compound? How can we rationalize our insistence that we must have wealth only via the earning of wages? You can have just the wages if that's all you want. As for me and my family, we will take the full wealth-creating benefit of our creative initiative and spiritual inspiration from both our Personal and Joint Capital Compounds!

America is now realizing fourteen to seventeen percent unemployment rates when all unemployed workers—those actively seeking work and those who have given up—are counted. The Progressive Socialists in general, and Obama in particular, are saying that we will see a return to more full employment by the end of 2010. Given the fact that everyone reading this book is now much better informed concerning wealth and its equitable distribution, do you really believe them? America, where is the hue and cry for wealth distribution that reflects the value and power of our Personal and Joint Capital Compounds?

I'm told that one should never ask a question without first knowing the answer. So, what do I think about the future of America and our unemployment numbers? Do I know the answer and will I risk giving it to you? Yes, I do know what will happen to America's unemployment numbers for this year 2010 and beyond and yes, I will tell you.

By the end of 2010, America's real unemployment numbers will move from fourteen to seventeen percent to sixteen to twenty percent. Beyond this year and on into the future, I predict that America's unemployment numbers will increase to 1930 levels of twenty to twenty-five percent. I believe this even though I am optimistic that America will be blessed by this book as well as other writers and speakers who are bringing this and similar messages to the grassroots. As we move forward, overcoming the many challenges that await us in transforming our economy from Progressive Laborism to true Capitalism, we will regain lower unemployment numbers. This is true because of the impact we will see from the equitable distribution of our Capital Compounds.

America is experiencing a great grassroots uprising as we will see this fall in the midterm elections. Americans will come quickly to understand the issues, even to embracing the equitable distribution of our complete Personal and Joint Capital Compounds. Having said that, why do I see unemployment crashing so badly?

I know that you already know the answer, but let me add what I can to assist your perspective.

We know that the Obama administration will not change course, even with overwhelming losses in these upcoming midterms. More stimulus money is in the pipeline as I write this book. There will continue to be more favoritism of organized labor. We are seeing this favoritism on our daily TV news programs, as we watch Toyota being shredded by bureaucrats, most of whom have received large sums of money from organized labor. Remember, Toyota does not use union labor.

We also know that the new housing sales are down more than even the critics thought it would be. We also see an increase in home foreclosures. What we don't see so easily is the foreclosures coming in the commercial real estate market. Only those involved in the REIT financial instruments will escape a major blow. This will add tremendous pressure to our already weak financial markets.

America's new coming grassroots revolution cannot stop these realities. These realities are in process: they are coming at us like an Abrams tank at full throttle. The impacts are happening even as I write this book. The effects of these impacts are only weeks, at most months, away. Some are here already. Sadly, they are destined to increase in scope as we move forward into this new decade.

The negative side of this new Health Care Bill, another example of Progressive Socialism's quest for power, will drag business further into economic chaos. Even stopping the process today would *not* keep the damaging effect from continuing for many years. Think of it in terms of a 747: now put the jet into a dive towards the ground. Then apply full power and raise the nose of the aircraft into a climbing position. The plane will eventually stop its decent and will start to climb back to a safe altitude. However, the weight and speed of the aircraft are a mathematical reality that dictates a certain predetermined point in the sky where it will start to regain altitude. If the maneuver is executed too close to the ground there is nothing that will save the airplane and all on board—it will crash!

Industries, large and small, are not stupid. America's small business owners, Marilyn and I included, are intelligent people who do not want to lose their businesses. With today's access to technology, the first place we all look to invest to improve our businesses, to expand our services and products, is at technology. This makes us all more efficient. Then, after we have time to absorb the impact of that investment, we will determine if we need to do more of that, or if we need to add a real live person to the payroll.

Unfortunately for all those Progressive Socialists who believe in laborism, the larger the company, the more they will turn *first* to technology to improve their company's efficiencies, productivity, and market share. They will turn to rehiring old layoffs and then looking at new hires *only* after they have maximized the efficiency of the newly applied technologies. However, it is not guaranteed that companies will automatically return to hiring people to rebuild and to expand. I personally believe that we will find new and better uses of technology, even additional uses for artificial intelligence to assist our forward movement ahead, without the old ways of using more and more employees.

You have already learned all of this from reading what I have been sharing with you thus far. You also have a good idea of what I am going to say next.

Yes, our future is not cast in stone. Yes, we will finally reap the benefits of our complete Personal and Joint Capital Compounds. The new leaders that we will elect in this year's midterm will seek out new answers to the old problems. The realities that we are learning through this book and other sources will reap their fruit and we will see creative steps taken to establish means of wealth distribution commensurate with our personal Capital Compound and that of our Joint Capital Compound. However, there is a price to pay.

Before we see the positive results of the future, we are going to go through a very difficult time. What we are suffering today is *not* a difficult time. What we have coming in the future *is* a difficult time.

Awake America, rebuild your God-given capitalist foundations, trust your faith in God, and put your shoulders to the tasks at hand: voting, talking to friends, seeking the truth, and committing it to memory. By doing this and doing it well we will regain the political power necessary to forge new financial instruments for equitable distribution of our intangible elements. First, we must regain the political advantage at all levels: local, state and federal. Party affiliation is secondary to character, integrity, honesty, transparency, and a commitment to protect our Constitution in its original form and application. Choose these men and women and we will have the power needed at all levels.

WHAT ARE THE DIFFERENT TAXES WE PAY?

While a complete list of all taxes we pay would not be practical or relevant to our discussion, a list of the most common taxes will help you visualize some of the ways our money makes its way into the hands of the government.

1. Individual income taxes (federal and also state for most states)

2. Corporate income taxes

3. Employment taxes (Social Security, Unemployment, and Disability)

4. Excise taxes, often referred to as the "hidden taxes"

5. Inheritance and gift taxes

6. Sales taxes

7. Property taxes

8. Personal property taxes

9. School taxes

10. Licenses and different permit fees

11. Inflation taxes, often referred to as the "silent taxes"

12. Welfare taxes—the redistribution of wealth through taxation

Although this list seems long and varied, it can be viewed from a different perspective that simplifies it considerably. For example, if we block together all of the taxes that are represented in the consumer goods, then we condense all but four of the list into the "added on" cost we pay when we buy something. All of the taxes except the inheritance and gift taxes, property taxes, personal taxes, and school taxes are added on to the price of the goods and services that we buy. Surprised? Most people don't realize all the ways taxes get passed on to the consumer, and we have only listed the more common ones.

For example, many workers feel some sort of satisfaction in knowing that large corporations have to pay an income tax, as though the corporation was somehow being punished just as they were. The truth is that the corporation or local business simply adds the cost of any taxes levied against them to the purchase price of the goods or services represented. As if that wasn't enough to discourage already tax-weary workers, the corporations more often than not do not have to pay the taxes that the consumer has been charged extra to cover.

You might think that I'm referring to some sort of illegal conduct on the part of the corporations, but I'm not. These are not sales taxes, but rather corporate income taxes, as well as other taxes passed on to them by suppliers of raw materials and semi-finished goods. The fact is that corporations have many ways—such as expanding their plant and equipment inventory and other expansions of their endeavors—that can qualify them to pay little or no taxes. As a result, the anticipated cost of the tax that was added to the item or service being produced is not required to be paid in the end. In such cases the consumer simply paid more than needed for what they received. However, it must be remembered that the business owners are only doing what is natural to protect their own interests.

Before we get too involved in examining who pays taxes, let's go on to our next question.

WHO OR WHAT CONTROLS TAXES?

The worldview and ideology of our leaders are always reflected in the goals of government. These goals, pro-Progressive laborism or pro-true Capitalism, will determine both the uses of taxes and the different means used to raise them. It is very naive to assume that government establishes its goals independent of the will of the people. A German philosopher once said, "The people of every society deserve the rulers that rule them." We might find fault with that to some degree, but when you think about it, there is a lot of truth in what he said. We might suffer under the wrong rulers for a time, but in the end it is up to the citizens to change the wrong and the bad for the good. If the citizens can't or won't, then they must suffer the consequences.

I remember with great consternation my feelings of helplessness during the last election cycle. I recall how, no matter what truth was revealed about Obama, even using his own words, those I spoke with could find no fault in him. Truly we have brought him to power and truly we will suffer the consequences, with no one to blame but ourselves.

We complain about lives, our problems, and the apparent lack of enough wealth to go around. We complain about taxes, wars, too much government, and many other displeasures of our life. The big question is how much effort do we put into trying to make changes? Or are we content to say, "The government is at fault, not me." I don't know about you, but to me that sounds a little bit like, "The devil made me do it!" You and your single shot voter cannon are absolutely important to both the success of America and the success of you, your children, and your grandchildren.

In a democratic society, the ultimate control of any action of the government—taxes or otherwise—is in the hands of the people. We do have the power to make changes. Presidents and members of Congress as a rule do not last long if they are not sensitive to the direction already established by the people. Before the 1800s most Americans were property owners, farmers, or skilled craftsmen with their own businesses. You couldn't even vote if you did not own property. Progressive laborism was the farthest thing from the people's minds. They wanted more land with opportunities on the frontier, and that's what they got. Even our inhumane acts against the Indian nations were carried out by the government primarily in answer to the will of the people. If it had not been for the sensitivity of our founding fathers and the many other great leaders, both black and white, who followed them and led a campaign against chattel slavery in the South, we might still be trying to make that inhumane system work.

As another example, one could say that President Roosevelt was not the sole author of his New Deal. Such ideas for social programs were already being voiced among labor leaders, and even big business was crying out for someone to do something. Keynes had laid down the theory of how to do it, the people were disillusioned with Laissez Faire, labor leaders and their memberships wanted a crack at seeing laborism actively supported by government, and so the New Deal was born. It wasn't so much President Roosevelt's New Deal, as it was *our* "New Deal." How do you like it?

The people, through the democratic process, give the power to the elected representatives to determine the ways and means by which government shall levy taxes. The elected legislative bodies, both on the state and the federal level, are the ones who present the bills and reforms that attempt to carry out the concerns the people have voiced from the home front.

As we have progressed from an agrarian economy of farmers supported by a few tradesmen to a complex, far-flung international industrial giant, the average voter has lost track of what it's all about.

"TAXES" YOU SAY?

> Who wants to be bothered about such matters? Just give me a secure job, lots of benefits, a fat check, a four day week, and a guaranteed retirement. Let somebody else worry about the taxes.

Sound familiar? Probably not in so many words, but the attitude is clear. It is that attitude that the politicians, labor leaders, and special interest groups tune into and it is that kind of government that we get in return. The ultimate answer to our original question (Who controls taxes?) is as simple as it is personally humbling: we the people control taxes! So, let's get informed and get about doing it.

WHO PAYS TAXES?

It's a common belief that everyone pays taxes, and indeed we can prove that they do. The frustration is that we don't all pay the same kinds of taxes, nor do we pay proportionately the same in relation to the amount of money we make. We pay many taxes other than income tax, yet it is income tax that gets the most media coverage. The sad truth is that forty-seven percent of all Americans do not pay any income taxes!

Our mixed economy, part Capitalistic and part laboristic, makes fair and equitable taxation very difficult to accomplish and very impractical to administer. Since Marilyn and I left the new capital acquisition industry for the small businesses industry, we established an accounting company for small businesses. Our services range from payables, payrolls, monthly statements, quarterly statements, sales taxes for different states, state income taxes, federal taxes and also IRS remediation services for those companies who have allowed themselves to fall into IRS default to the point of having their accounts levied and or seized. The breadth and depth of all that we do and have experienced has given us exceptional training in these matters. So when we tell you these things, we are speaking from very close experience to the problems.

Those who are only concerned with their take-home pay look at their tax deductions and say the rich should pay more. The irony of that cry for justice is the fact that, only a few decades ago, the standard of living that a worker now enjoys was reserved only for the rich! So, maybe the earlier generations have been granted their wish, and the "rich" are paying more.

More importantly, perhaps being rich is a matter of perspective. If you have more money than I do, then to me you might seem "rich." If someone else has less money than I do, I might consider that person to be poor, and vice versa. Such cries for equitable tax treatment based on a misconception of who is rich and who is poor do not treat the subject with justice. We must look deeper into the forces that have shaped our tax structures and that continue to propel them forward.

At the same time the worker is complaining about overtaxation, legislators are trying to figure out ways to get people to part with more and more of their earnings so that they (who don't believe in Capitalism to begin with) can "invest" it in another nonproductive government program. This is the Progressive Socialist's version of Capitalism—the part that the government is trying to encourage to keep up with the objectives of Full Employment." After all, isn't that what we have all bargained for—Full Employment?

Well, it's not so easy to get people to part with their money to invest in new things, or even to expand existing projects. True, there are some people who will invest without the encouragement of government or anyone else, but that is not enough when you are saddled with Progressive laborism and you have everything tied to the concept of Full Employment. If the unemployment rate increases, consumer consumption goes down, demand goes down, and less tax money will be raised through sales and income tax than before. Benefits can't get paid, and welfare rolls start to swell. Labor leaders and politicians alike begin to worry. The Progressive Socialists made a deal remember—Full Employment—it's the name of their laborism game.

So what can be done to encourage the flow of enough investment money to keep the capitalistic side of our mix strong enough to maximize employment and keep up consumer buying? Legislators don't have all of the answers to that one, but the so-called Conservatives have learned to give tax breaks and tax incentives in return for investments. At the same time that this idea is effective in increasing the amount of investment capital that is possible to raise each year, it also creates its own kind of problems.

Anytime that an exception can be made for one class of people and not another, you create an environment in which abuses are likely to take place. Those who have enough money to be able to cover their living costs and still have extra to invest in those areas where our legislators have offered tax breaks pay fewer taxes and make additional money on the investment. It is a nice way to avoid the tax burden, if you can afford it. "Unfair!" I hear someone cry, and on the surface it might seem to be so. But under our present system of a mixed economy, part Capitalistic and part laboristic, it's not only *not unfair*, but absolutely necessary. The "unfair" part disappears when you consider that the investors are willing to assume the risk of the investment, and that, without them, unemployment would rise.

What *is* unfair is the fact that everyone has not been compensated for the total contributions that they have been making over the years with their Personal Capital Compound. If everyone was receiving compensation for their total Capital Compound and its synergistic contributions, in both wages and equity shares of ownership, such unfairness could be eliminated. Until such time, we will continue to be saddled with only a *semi-Capitalistic* economy, one that will never survive the labor absorbing power of technology!

In one sense of the word, you can actually say that paying income taxes is a penalty for not investing enough money into the Capitalistic side of our mixture. Take a good look at your Form 1040, the income tax report form, and in particular take note of all of the allowable deductions that are possible. How many of them can you honestly say do not have some relation to, or bearing on a Capitalistic function, either directly or indirectly. Even your own personal deduction is allowed because you are alive and a consumer. No dead people can be claimed—they don't consume anymore!

Our problem is simple: we are trying to burn the stick at both ends and avoid getting burned. We must make exceptions to the tax rules in order to encourage investment to support the Capitalistic side of our mix, and we have to support the ever growing numbers of workers with more and more jobs that pay more and more money.

At the same time all of this is going on, someone has forgotten to tell our two intangible elements to take a long vacation. They don't seem to know about our laborism and semi-Capitalism mix. They are busy trying to perfect the fifth and sixth generation computers and artificial intelligence so that men and women can be replaced by more efficient and more reliable substitutes!

Thus, labor leaders rush to the forefront in a show of deep concern for the American wage earner and make a passionate plea for—what else—job security! Let's be reasonable: who's kidding who?

One of the structures we have set in motion to keep this dilemma alive is our corporate income tax structure. This little piece of artwork is one of the main reasons that capital is so hard to raise for investments in plant and equipment. It is this structure that says that corporations must pay income taxes when they declare a profit. As we have already mentioned, all the corporation does is pass that tax cost on to the consumer. However, before the corporation can declare a dividend, they must first pay taxes on the profits made, then the shareholder pays a tax on the dividend received, which makes the second time the same money has been taxed. This is double taxation. This would not be so bad if this system of laws also required the corporations to distribute all of their profits each year to their rightful owners, the shareholders.

When I presented to you the unique qualities of the REIT—the Real Estate Investment Trust—you were seeing the very latest in capital instrument design that performs just exactly as the Joint Capital Compound provides. It is possible to restructure regular C corporations, LLCs and Sub-S capital instruments to perform with the same distribution efficiency as the REIT, but we have not begun to create these new designs. Such designs require a major move away from laborism as the only true source of wealth distribution for the average American. We must first reject, with great determination and a clear focus, all aspects of Progressive Socialist's union-supported laborism!

But, alas, we are more laboristic than Capitalistic and therefore those trusted political representatives who are listening to our hue and cry from the home front for "more cash upfront" do not see any need to make changes. In the absence of doing what is right and just for the owners of the shares, they make it preferable for the corporations to retain large earnings—profits—that the board of directors decides not to declare. These retained earnings make up the bulk of the new capital that is used for plant and equipment expansion, and at the same time discourages others from being so quick to buy stocks in which they will have little or no control.

The only practical, equitable and efficient way to increase tax roll revenues is to establish a *flat tax* on income! A flat tax of twelve to fifteen percent for everyone without the long complicated formula for deductions would collect more than enough money for all our federal government's constitutionally authorized involvement in the execution of its sworn duties and responsibilities. Just this one adjustment away from laborism and towards true Capitalism would invigorate our American economy to new heights of power and success, with lower prices for products and services. I predict that we will do this and more. We have no choice: we either embrace the realities of true Capitalism or we will lose our beloved America. When we make the change from Progressive laborism to true Capitalism, we will reignite the hope of the world for true economic reform and we will lead the world into an era of true Capitalism never before experienced in history!

Dr. Martin Luther King said, "Injustice anywhere is a threat to justice everywhere." Certainly we can see that our attempt to support laborism's ideal of Full Employment, and at the same time encourage more plant and equipment expansion with its accompanying technological advances, creates injustices everywhere. We must come quickly to grips with reality if we are to forestall future tragedies of even more serious consequences than the Great Depression.

Unfortunately we are already too late to divert such an economic experience—as I have clearly revealed, we are heading there at full throttle! We cannot stop this plunge into economic chaos. We must stand up and put our shoulders to the tasks ahead of changing our leaders and rebuilding our God-given capitalist foundations. Our future is not cast in stone! We can rebuild and lead the way to great success!

WELFARE TAXES

Welfare is a type of "charity"—a public charity—one to which everyone contributes, even those who receive support from welfare. In a society such as ours that uses laborism as a primary means of distribution, such public charity is a must: and not only the public charity of welfare, but the many private charities as well. Our local churches, societies of different types, home mission boards, rescue missions, and thousands of nonprofit organizations that endeavor to help a broad "underprivileged" segment of our society are all needed to fill in the gaps in our ineffective and inequitable distribution process.

The ideal of laborism is Full Employment all the time. However, the reality of Progressive laborism is that it accepts unemployment as a necessary evil that must be dealt with as the semi-Capitalist side of our mix makes adjustments for market changes and fluctuations. The reality of this was not ignored by Marx and Engels. It is this very weakness in the semi-Capitalism and laborism mix that caused Marx and Engels to believe that the only solution was to turn all private property over to the state and forget the whole mess. In this role the state was to assume the position of national employer and national charity dispenser all in one. Why does this sound so much like Obama and his Progressive Socialists?

Supporting the charity side of our mix requires an enormous amount of diverted taxes and the cost of all of this is truly a tax. Those dollars that are directly acquired by the various government agencies, both federal and state, are represented in the increased cost we consumers must pay for our goods and services. The dollars that we give to all of the other charitable organizations are dollars that could have been put into owning capital instruments of production. Without the need for such expenditures to take care of the unemployed and the dropouts at the lower end of the labor pool, the cost of caring for the *truly needy* would be minimal.

The cost of government sponsored *food programs alone* throughout the United States comes to over $125 billion annually. This does not even take into consideration additional billions that are represented in all of the private endeavors supporting food distribution centers all across the country.

A family must face these same realities. A father and mother can "bring home the bacon" to the hungry youngsters, but they cannot avoid the discipline of budgeting this income.

Every budget must have disciplines. If disciplines and guidelines are not structured and maintained, it is possible to "find enough needs"—food, housing, education, entertainment, etc—to outspend the healthiest of incomes. Expressions like "too much month at the end of the money" only emphasize the mindset of many families enamored with consumerism. The majority of Americans are simply following our Progressive Socialist government's example of deficit spending.

Our country is faced with ever increasing personal bankruptcies, resulting in painful disruption of family harmony, with many breakups in relationships that end in divorce. These broken families further strain the fiber of our culture and economy.

America, along with all non-Communist countries, has accepted laborism as a modification of Marx's Labor Theory of Value. As a consequence, we have structured our tax laws, corporate laws, inheritance laws, and all of our social service support systems and institutions around this basic premise. Laborism has fostered such models as the "trickle-down" theory. Major owners of equity in the instruments of production, and even some labor leaders, have used this theory to justify the large concentrations of capital ownership in the hands of only the few.

Also in the wake of our laborism we have created multitudes of the forgotten that we call welfare recipients and hardcore unemployed. This has created a lower class as though it were some kind of unavoidable disease that was not our fault. Poverty brokers cry out for more government grants and programs to further enslave the masses under their control. Black Americans suffer the most from their own kind of Progressive Socialist black leaders. Where are the millions that have been "helped" by such programs at the beginning of the New Deal, the Great Society and the War on Poverty? Have they not multiplied into yet more millions until, now, we have cities crammed full of third and fourth generation welfare families?

This is the plight of our Black Americans under the false claims of many of their most visible and outspoken representatives such as Reverend Al Sharpton, Minister Louis Farrakhan, Reverend Jesse Jackson and many less visible. All of these so-called friends of the Blacks spout the same litany of lies and half-truths to gain their objective to control as many Black Americans as possible. There is another side to tell even more representative of the failure of laborism than Black Americans. It is a little-known fact that there are many more White Americans on welfare than any other single group. They have fallen for the same lies and half-truths without the celebrity style spokesmen of their Black counterparts!

Until the American worker understands the value they represent in their total Personal and Joint Capital Compound, they will continue to pay the ever increasing welfare tax. This tax is not only costing us the loss of precious resources in the form of cash, but it is also costing us billions in the loss of the production power of the Capital Compound of everyone now on welfare. Obama has now added the new Health Care Bill to that cost.

This abominable welfare tax has a solution: it starts by dealing with the problem and not by turning millions of people into worse than chattel slaves. Slaves at least have dignity and pride in the work they accomplish. Welfare dependents are like condemned people with a life sentence. They have no goals, no purposes—their only concern is to exist. The only differences are that their prison walls are the borders of their ghetto and they can procreate.

The welfare tax and all of our benefit programs including Social Security, major medical, disability insurance, unemployment, and even inflation, are simply the tools of Progressive Socialists to redistribute wealth through taxation and to gain vast amounts of power in the process. This further supports the concepts of laborism and weakens the capital instrument base of our economy. These billions of dollars that are shifted from our capital instruments of production (research and development and marketing efforts) cause an increased strain on the economy at large and reduce our national and international competitiveness!

SUMMARY

Now that we know more of the nitty gritty about taxes—who pays them, who controls them, the good ones and the bad ones—how do you feel about taxes? Not too peachy keen, I imagine. The one thing that will make you feel not only better but exceedingly satisfied with yourself is the success you will realize through the process of America's transition from Progressive laborism to true Capitalism. Not only will our taxes go down, the taxes we *do* pay will go for needs and issues that are part and parcel of our beloved America's infrastructure: security at home and abroad.

And yes, they will provide for those souls who still remain of our last generation of welfare recipients. These are those who have no hope now of anything but subsistence living until they die and are buried. The New Deal, the Great Society and the War on Poverty—the hallmarks of Progressive Socialism and laborism—robbed them of their dignity, their opportunity to know their true value, and their hope for a successful experience of the pursuit of happiness. God will repay those responsible!

Chapter Thirteen

Our Future Is Not Cast in Stone

Often when we are challenged with stress and difficulties we can become excited about what might be if things were different. The good news is that things can be different. In fact they are going to be different, whether we want them to or not. The Progressive Socialists will never stop in their efforts to bring our beloved America into a Communist state with total control of everything, including you and me and our families!

The question is not if things are going to change. The question is what kind of change are we going to have? Our Personal and Joint Capital Compound works and produces wealth—it will produce wealth for whoever controls it. If our future is to be a Communist country, then the state will own and control it. However, if we repent of our sins of inactivity and lack of involvement in the political process, reeducate ourselves on the details of our true value as capital instruments for creating new wealth via our Personal and Joint Capital Compound, then our future is filled with sunshine, white clouds and much success within our lifetimes and for all of our grandchildren for many future generations.

WHAT ALTERNATIVES DO WE HAVE?

Alternatives are limited only by our vision and concept of who we are. If we are only muscle and bone, and all wealth as we know it comes only from the fruit of labor, then the laborism we've got is all there *is*. The Communist states should be bailing us out of our problems any day now. For surely they have practiced their belief and commitment in the Marxist Labor Theory of Value to the limit. It should be the Communist states who are riding the crest of the high technology wave. All of the world's greatest and most advanced discoveries should be flowing from Russia, China, Cuba, and Venezuela.

Instead, what do we see? We see those countries that have embraced such a limited vision and concept of who they are coming to the non-Communist countries and asking for loans and credit to help sustain their losing proposition. Then, to add insult to injury, we see them using all forms of devious means to buy or steal the high technology that seems to be always just beyond their reach.

Most alarming of all is that now in 2010 we are going to the largest Communist country still functioning in the world today. China, with her Communist style government and Capitalist style economy, now has the fastest growing economy in the world. The Chinese economy is not a true Capitalist economy and they do not recognize the personal Capital Compound and the Joint Capital Compound as a reality. They have no desire or apparent need to be concerned about the equitable distribution of the productivity of these two compounds.

China is many years away from the same kind of economic collapse that we and the rest of the Western world are now experiencing. The reason is very simple: masses of available labor, willing and able to work for a little as necessary just to get any kind of an income. Therefore, China has the luxury of working both ends to the middle, their centralized Communist styled government and their laboristic version of Capitalism. We on the other hand have run out the laborism clock. We are faced with the total collapse of our laborism based semi-Capitalistic economy. We can no longer create enough good jobs to overcome the efficiency and power of the intangible elements of our Personal and Joint Capital Compounds. So what are we doing?

Sadly, we are going to China to fund our outlandish drive to buy our way out of this dilemma. Obama and his cadre of Progressive

Socialists are working overtime to create new ways to spend money—all rationalized as "saving" our laboristic, semi-Capitalistic economy—to create new jobs, to create new green industries, to expand new Socialist services such as health care, Medicare and Medicaid and to at the same time *reduce* the federal debt! Only blind, deaf, and dumb Progressive ideologues could possibly see even one small ray of daylight in such a boondoggle! Tighten your seatbelts and stand by for crash landing!

NEW GOALS ARE NEEDED

The Capital Compound Theory of Value is not just a theory, it is an expression of productive dynamism which works all the time, everywhere there is life, whether the participants are aware of it or not. It does not require the acceptance or acknowledgement of the life it serves. It is a gift from God for all humanity. Our challenge is not to make applications of a theory, but rather to embrace the reality of its existence and allow it to bless us. It has been given to us for blessing not for cursing. It only curses us when we deny that it exists and inasmuch as we do not know the full limits of its capability. Its reality and its power to create new wealth is evident all around us. The three elements of our Capital Compound—labor, creative initiative, and spiritual inspiration—give us an opportunity to establish new goals. New goals will require that we adopt new methods and structure new approaches to reach them. If our goals as a country— and as statesmen, scientists, educators, businesspeople, or working people— had not included going to the moon, we would not have developed the technology to travel in space.

The Labor Theory of Value and laborism, our own Progressive Socialist modification of it, are dead. The proponents of it are simply blind, deaf, and dumb. The Russian model of total central control was a cadaver from the beginning, and would never have emerged from its Frankenstein laboratory if the West had not pumped hundreds of billions of dollars worth of cash and technology into it to make it appear alive.

Our own laborism modification only appears to have life because we have it hooked up to a life support system called Progressive

Socialism, tied to the general fund of our federal government exercising its declared right to manage the "general welfare of the people." We have accepted this contrivance as a representation of Capitalism, but the truth is we have *not* practiced true Capitalism for over a hundred and fifty years.

Assuming we can accept the reality of our Personal Capital Compound, our goals will automatically change from trying to pump blood into a lifeless cadaver to cultivating the productive, life-giving powers of our Capital Compound itself.

Our suggestions for possible alternatives certainly are not intended to represent an exhaustive study. The total work of refashioning our economy, with everyone sharing in the ownership of capital instruments of production, is a task in which we all must take an active part!

WE CAN EDUCATE OURSELVES

First of all *workers must learn who they are and understand the value of their Capital Compound.* Industry leaders, labor leaders, and politicians are waiting for the American worker to step forward and be counted. It's not the trickle-down that counts, it's the rising flood that will make the difference. A rising flood floats all boats!

WE CAN CHANGE OUR STATUS

The "rising flood" will be brought about by those millions of workers with a new vision and concept of who they are and what they are worth. They will exert the kind of grassroots pressure that will be needed to make the necessary basic changes—changes such as the status of our economic rights as citizens. Our constitution provides that all men and women should be free to vote, with the freedom to protect our rights to the "pursuit of happiness." This being true, we cannot be denied the right to possess and own as private property the capital instruments necessary to secure a standard of living based on equity and not limited to labor. In times past, the right to bargain for a wage might have seemed to many to be enough—they never took notice that they were giving up their most valuable contributions, their creative initiative and spiritual inspiration!

We need to surround the economic status of the individual capitalist or stockholder with legal protections and privileges similar to those we have given the political status of the individual citizen. The individual capitalist, like the citizen, is a person who needs to be able to exert legal power in the control of his own economic affairs. From such changes the "capitalist citizen" will emerge. Interestingly enough, corporate law already does much of this for corporations, giving them personality and rights even as people.

WE CAN HONOR THE OWNERSHIP OF THE SHARE HOLDERS

The changes necessary to make it mandatory for corporations and businesses to distribute all of their profits each and every year to their stockholders and investors would take priority over laboristic legislation. All new capital should be raised each year from the open market. This would give the stock market a method by which to measure more accurately the performance of companies, and the gambling aspect of the stock market would be reduced considerably, if not entirely eliminated. Citizens would feel good about buying stocks for their income potential rather than to play with them as gambling chips.

WE CAN ELIMINATE DOUBLE TAXATION OF CORPORATE PROFITS

Corporate income taxes should also be eliminated at the same time. This would stop the useless double taxation of profits, and reduce the expensive accounting procedures some corporations adopt to justify not showing a profit.

The Real Estate Investment Trust, REIT, already does this and shows the way.

Neither would they have any further need to calculate the tax cost into their products or services. Real costs of goods and services would become more visible, and their unit production cost would decrease. Thus we would see an immediate improvement in national and international competition, and a growth in GDP with a lowering of unemployment. This combination will give us enough time to fully establish the new distribution reality of our Capital Compounds.

This rise in productivity and expanded GDP will create a tax base with longevity and volume. Under true Capitalism there is no such thing as insufficient tax revenue. The general fund of the federal government would always be in excess of the expenditures required for Constitutional government involvement in defense, infrastructure, judicial, and executive duties. There is also a direct increase in the income of all Americans as they are owners of corporate shares and capital instruments.

Under our current laboristic, semi-Capitalistic economy, most of the shareholders earnings have been held back as "retained earnings" by the decisions of the board of directors of the corporations. This is done to increase the holdings of the corporation and at the same time to avoid the corporate income tax, which for the corporate directors is a natural response to our present tax and laboristic distribution methods. Consequently, under the new Personal and Joint Capital Compounds distribution of wealth via equities, the shareholders would be enjoying new income that they was formerly denied. Along with the new income they would also pay their *flat tax* share of taxes, albeit on a larger gross income. It is very possible that they would want to reinvest their profits into more shares of the same corporation or private company, and they could do that through the open market. However, if they did not want to reinvest it would be their decision and not the board of directors.

WE CAN ELIMINATE INHERITANCE TAXES

Inheritance taxes need to be eliminated, as these are only needed where we are striving to support a laborism policy of distribution. The right of a person to own property is clear in our constitution. As an owner of property, the right to control who gets said property in the event of their death is implied. Such taxes only serve to reduce the ability of families to accumulate a truly meaningful amount of capital instrument assets. When distribution through ownership in the instruments of production is the basic concept and method of equitable distribution of new wealth, then such changes make sense. Also, as we move closer to the day when artificial intelligence becomes a practical tool for replacing most of our present human effort, ownership by everyone in such instruments of production will be a must.

WE CAN REDUCE MONOPOLISTIC TENDENCIES

Efforts to protect the individual Capitalist and stockholder from the unfairness of monopolization are needed. We do not encourage or believe in market monopolization, and neither should we support the monopolization of participation in the ownership of our country's instruments of production.

WE CAN REDUCE THE OVER-CENTRALIZATION OF PRIVATE CAPITAL

Some determination of the maximum amount of ownership allowable in the instruments of production would be needed as we moved from a laboristic taxation and distribution system to a Capital Compound system of taxation and distribution.

Obviously the upper limits should be flexible, for several reasons. It is always possible that the cost of living might increase somewhat, even in a Capital Compound economy. There should never be a limit to the amount of capital instruments that one could own. Success is not a viable target for taxation. We currently punish success through taxes and other means, but that is because we are working under the illusion of Progressive Socialist laborism. Additional purchases of the instruments of production by the same family is not bad, it only increases the gross income of the family and subsequently their tax bill. This would not interfere in any way with the normal "pursuits of happiness" of our citizens, but it would increase the tax base.

Under the normal competition reflected in a healthy marketplace—based on true Capitalism with equitable distribution of equity based wealth—there would be *no* unfair accumulations of capital.

THOSE AT THE TOP AND THOSE AT THE BOTTOM

For those who already own large amounts of wealth represented in the instruments of production, their future will be secured from the ultimate tyranny of the state. Without the diffusion of ownership in the instruments of production, millions of displaced workers in the near future will have no other recourse but to cast their vote for massive takeovers by the federal government. Unrealistic greed and pride are the only things that would cause those who now have large fortunes to feel unthreatened.

Those at the bottom of our economic chain—those that now are supported by welfare and private charity—we would have to continue to support. All able-bodied persons in those ranks could receive additional training and return to be productive citizens. The complete re-education of this group and those that are only marginally involved in the economy will require more than training in hand tools and simple labor tasks. These kinds of jobs are disappearing into the hands of robots and less expensive labor in other countries. This group, as well as all America, needs to expand their understanding of entrepreneurism, and the advantages of owning one's own business. Small business opportunities are the crucible of the most productive aspects of the elements of our Capital Compounds. This area of industry will grow exponentially under true Capitalism.

As a large number of major corporations and companies secure work contracts with their families of employees based on the total distribution of their Joint Capital Compound, company competitiveness would increase and unemployment would decrease. A gradual decrease in the welfare and charity rolls would ultimately put an end to that kind of nonexistence once and for all, except the most severely inept and the totally disabled.

CHOOSING FREEDOM OR SERFDOM

Candidly, all of the policies cannot be changed at once, and there may be others which will take on greater import as many committed and creative people get involved in the process. But we must get the ball rolling and attract the best minds to structure future guidelines. Even now Congress is adjusting policies first one way and then the next, grasping at straws, hoping to stumble onto something that will work. Until they get their head out of the Progressive Socialist's honey bucket, they will never come close to the answer!

We've already concluded that going on with the same escalating indebtedness to support Full Employment for all is an exercise in frustration. Turning the tide of government and corporate policy, however, will take a major change in contemporary thought and understanding. Only a momentous wave of demand from "we the people" can turn the forces before the thrust of our present laboristic momentum sweeps us into economic and social disaster, all the way to *serfdom!*

Hall of Fame basketball player, Larry Bird, once said, "A winner is someone who recognizes his God-given talents, works his tail off to develop them into skills, and uses these skills to accomplish his goals."

Does it seem like too much work to develop those skills you know you don't have but need?

If you are more technical, you have great *hard* skills; *if you are* more relational, you have great *soft* skills. It will take work to develop the one you don't have. That's understandable—some people call that "pain management." Your mind tells you that in order to get better at your job or *to* be able to influence others to be better, you need *soft* skills that are just too *difficult* to develop. So you continue to do what you do, hoping to get a different result (Einst*ein*'s definition of insanity).

Have you considered the reward for the work in developing those needed skills? Consider all the ways you would overcome, move forward, be respected, and gain results. We need the skill of voting and the skill of encouraging others to vote. We need the skill of opening our minds to new proven concepts that will equip us to be and do as we have always dreamed of being and doing.

Have you taken time to consider the gains? And what do you stand to lose by not developing the needed skills listed above?

Taking into consideration that it takes no effort at all to keep on doing what we've always done, it should come as no surprise to us that we keep on getting what we've always gotten.

SUMMARY

We have a small window of opportunity to take action. We are all aware that we need to know some things in order to be able to ask the right questions of our political hopefuls and those who are already in power. The reality of our Personal and Joint Capital Compounds is a fact of our life, whether we know it and recognize it or not! My heartfelt prayer is that all of us will do all that we can where we live and with those that we know, to spread the word about this book and its message of answers and hopes for a great future blessed by God!

God bless you personally, your family, your business and all those you love and serve!

Zester and Marilyn Hatfield

Acknowledgments

Although it would be impossible to acknowledge all of the many individuals whose encouragement and input have contributed to the publication of this book, we do want to make note with sincere gratitude the contribution of the following individuals and companies:

First of all, to the countless number of wage earners whose economic struggles we have accompanied—resolving some and helpless to resolve others in our consulting practice, and for their open communication in giving us insight into their need.

To the many small business owners that we have known and that we have served for many years.

To the many union local offices in Northern California and their business representatives who have referred members to us and who have given us vital information on the problems facing union growth.

To our public library system which is often taken for granted in this country, but which provides a wealth of information on the history of our economy.

To Mr. Richard Haley, Palo Alto, CA., friend and consultant, whose knowledge of corporate structure has given much needed insight, and whose encouragement has been priceless.

To Dr. Orv Owens, friend and corporate psychologist, whose insight into what motivates people has been a most valuable resource. In promoting the best interest of the wage earner and maximizing his potential for economic freedom and peace of mind, Dr. Owen's courses have proved extremely beneficial, as well as his personal encouragement to pursue this course of action.

To Louis Kelso, San Francisco, CA., author of *The Capitalist Manifesto;* father of the ESOP, *Employee Securities Option Plan,* has led the way for employees to bargain for "more than wages."

To Dr. John Perkins, a pioneer in Black Christian Community Development, whose application of the capital principle has taught and provided economic freedom to the poor and underprivileged, and whose dedication has been a personal source of inspiration.

We acknowledge Josue Lopez, Juarez, Chih. Mexico, our lifetime friend, and the administrator of Immanuel Children's Home. His personal encouragement and insight into the economic desperation of the Latin American countries is of critical importance.

To the late Rev. Loran Biggs, Senior Pastor of the Pacific Ave. Christian Church of Santa Rosa, California, life-long friend and family pastor whose personal encouragement often meant the difference between continuing on or giving up.

To E. B. Rich, Sub-District Director of United Steel Workers of America in Fairfield, Alabama; a Union leader who not only understands the needs of his people, but whose insight into industry and market demands made possible a viable contract with U.S. Steel in building the most modern seamless pipe mill worldwide, providing present jobs where a continued shutdown was the only alternative and future jobs in his area when other mills will be faced with obsolescence. His input has not only been valuable for the book, but his unsung example has led the way in his industry.

To D. Sloan Hill, Public Affairs Manager at U.S. Steel Plant in Fairfield, Alabama, for his time and energy in sharing with us his corporation's contribution toward maintaining its place in the market and providing jobs and benefits to thousands of wage earners. Mr. Hill also contributed the photographs of the seamless pipe mill.

To Donald Clay, General Manager, Public Affairs, and Warren Hull of U.S. Steel's Pittsburgh, Pennsylvania, office for their contribution.

To Mr. Jackson M. Saunders of GMF Robotics, to John Grix, Manager, Public Affairs, of the Warren, Michigan offices and to Thomas F. Macan, Manager, Public Affairs, at the Detroit offices for their part in making possible our tour, the many photographs, and videotape of General Motors' most modern plant in Orion, Michigan.

For the contribution from each of the following corporations of vital material and data, we express our heartfelt gratitude:

Mr. D. P. Crew and Mr. Brian R. Gareau, Public Information Representatives, Caterpillar Tractor Company, Peoria, Illinois.

Ms. Juliet A. McGhie, Public Affairs Representative, Exxon Corporation, New York, New York.

Mr. William H. Jones and Mr. Dale E. Basye, Manager, Corporate Communications, Chevron Corporation, San Francisco, California.

Mr. Leon R. Brodeur, Vice Chairman, and Michael E. Fay, Manager, Community Responsibilities, Firestone Tire and Rubber Company, Akron, Ohio.

Mr. Lane Webster, Public Relations Services and Mr. David Kirby, Director, Hewlett-Packard Corporation, Palo Alto, California.

To Jim Dunham, retired math teacher, Nixa, MO, our friend and loyal brother-in-law, for his dedicated editing and content suggestions.

To Paul Rule, Captain U.S. Air, a close friend and neighbor in Timberon, New Mexico. Paul also designed the graphic for The Capital Compound Theory of Value.

To Mick Garrett, President of BuildBlock, LLC, Oklahoma, OK, a close friend and business associate, for his insightful content suggestions.

To Le-En Tia, Director of the 10500 Ministry Program, Chicago, IL, a close friend and ministry associate.

To Rio Lynne García, T.E.A. Party Coordinator, Tempe, AZ, a close friend and business associate, for her review and great Support.

To Todd Peterson, VP and IT Manager for Desert Communications Inc., a close friend and business associate, for his insightful content and structuring input.

To our wonderful children who love and encouragement is priceless. To our twenty four grand children who keep us believing in God's overcoming grace. To our parents, Zester and Ruth Hatfield, and Wayne and Betty Jones for being the kind of parents who taught us to know, "By God's grace, you can if you think you can."

To all our friends and associates, and to our beloved country that dreams big dreams and that believes God is with America and will bless her now and for evermore. We appreciate the opportunity to contribute.

Others Comment
"Progressivism: Our Road
To Serfdom"

"A very timely book; outlines plan for a new era of labor/management cooperation. Addresses a growing problem of many U.S. corporations - how to remain competitive, treat employees fairly, and prepare for the next generation of technology."

Will Pilcher Economist *Levi Strauss & Co.*

"This new book should be required reading by all corporate executives and government officials who control the lives of people through policy decisions ... a classic in the same mold as *Wealth of Nations by* Adam Smith."

James W. Kunz President *National Research Center Colleges & University Adm.*

"Enlightening futuristic approach, to an inevitable employment marketplace! Profitable reading and a blueprint for your career planning!"

Louis Kram, CPC President *Best Personnel*

"Reading this book is a must for every wage-earner. It will convince organized labor that the shifts from heavy manufacturing toward high-tech and service industries need not be antithetical to the workers' long-term interests ... it seems Labor has a lot of self-education to complete before it has properly weeded its own garden."

Richard H. Haley Corporate Consultant *Silicon Valley, California*

"The Hatfield's have researched answers to world financial problems in a practical and creative way; highly recommended reading!"

Dr. Orv Owens Psychologist and Corporate Consultant *Washington, D.C.*

"I have known the Zester & Marilyn Hatfield for 5 years, and have seen the benefit of their research for the working man and woman. As a leader in the labor movement, I can wholeheartedly recommend their work to every American."

Jim Cupp Local 1235 Business Representative & Recording Secretary *San Francisco, California*

Rio Lynne Garcia, T.E.A. Party Coordinator, Tempe, Arizona, "Beyond Wonderful!"

Contact Information

Hatfield & Company, Inc.

Accounting and Tax Services

IRS Remediation Services

Marilyn Hatfield

Email: marilyn9742@gmail.com

Business Consulting

Equity Sharing Financial Instruments

Management Training for Capital Compound Instruments-CCI

Department & Division Manager Training for CCI

Employee Training for CCI

T.E.A. Party Presentations and All Other Public Speaking

For

Zester Hatfield

Contact Rio Lynne Garcia, T.E.A. Party Coordinator

Email: Business Cell: 480-784-7948

Email: riogarcia@live.com

Book Purchasing

www.progressivesocialism.com